Vegetable Production: Contemporary Techniques

Vegetable Production: Contemporary Techniques

Edited by Pearl Hunter

SYRAWOOD
PUBLISHING HOUSE

New York

Published by Syrawood Publishing House,
750 Third Avenue, 9ᵗʰ Floor,
New York, NY 10017, USA
www.syrawoodpublishinghouse.com

Vegetable Production: Contemporary Techniques
Edited by Pearl Hunter

International Standard Book Number: 978-1-68286-720-4 (Hardback)

Cataloging-in-Publication Data

Vegetable production : contemporary techniques / edited by Pearl Hunter.
    p. cm.
Includes bibliographical references and index.
ISBN 978-1-68286-720-4
1. Vegetables. 2. Vegetable gardening. 3. Horticulture. 4. Seed technology. I. Hunter, Pearl.
SB320.9 .V44 2019
635--dc23

TABLE OF CONTENTS

PREFACE

Vegetables play a significant role in human nutrition. Most vegetables have low fat content but are replete with essential vitamins, minerals and dietary fibers. Vegetable production includes all agricultural practices of soil preparation, weed removal, irrigation, harvesting and storage. Different soil types are suitable for different vegetables. Vegetable production is also affected by seasons, rainfall, temperature, etc. In modern times, cultivation is witnessing increased mechanization to reduce human labor and increase productivity. Raised bed gardening and hydroponics are some of the modern technologies of increasing vegetable production significantly. This book aims to provide detailed information about some advanced concepts and theories related to vegetable production. It also aims to shed light on some of the unexplored aspects and the recent researches in this field. Agriculturists, horticulturists and students actively engaged in this area of study will immensely benefit from this book.

All of the data presented henceforth, was collaborated in the wake of recent advancements in the field. The aim of this book is to present the diversified developments from across the globe in a comprehensible manner. The opinions expressed in each chapter belong solely to the contributing authors. Their interpretations of the topics are the integral part of this book, which I have carefully compiled for a better understanding of the readers.

At the end, I would like to thank all those who dedicated their time and efforts for the successful completion of this book. I also wish to convey my gratitude towards my friends and family who supported me at every step.

Editor

NEW PROCESSING TOMATO VARIETIES OBTAINED AT V.R.D.S. BUZĂU

Costel VÎNĂTORU[1], Bianca ZAMFIR[1], Camelia BRATU[1], Victor LĂCĂTUȘ[2], Luminița CÂRSTEA[2]

[1]Vegetable Research and Development Station Buzău, No. 23, Mesteacănului Street, zip code 120024, Buzău, Romania
[2] Research and Development Institute for Vegetable and Flower Growing Vidra, 22 Calea Bucuresti Street, Bucharest, Romania
Corresponding author email: costel_vinatoru@yahoo.com

Abstract

Researchers for improving tomatoes at V.R.D.S.Buzau have been since its beginning, 1957. Here were obtained for the first time in Romania, the tomato hybrid seed under Bulgarian license, the 10 x Bizon famous hybrid. Over time the new valuable, with well-defined genetic constitution varieties were created which currently occupy significant surfaces in culture. Among these are: Buzau 22, Buzau 47, Buzau 1600. Studies undertaken by V.R.D.S.Buzau about the Romanian varieties of tomatoes have spotlighted that old varieties of tomatoes were patented as mixed, with destination for fresh consumption and industrialization. At the same time it was found that these varieties do not always meet the appropriate parameters for both destinations, so after the year 1990 researches were undertaken by the Improvement Laboratory in an intensive system, with the aim of obtaining creations strictly specialized, according to destination. The research started with the establishment of a solid germplasm database, followed then by its knowledge and use in the improvement process. The unit currently has a total of over 1000 genotypes from this species in various stages of improvement. The main objectives for improvement of which were obtained new creations were: productivity, quality of fruit in accordance with the requirements of the processors, the dry substance content, low acidity, pigmentation, the content of lycopene, sugar-acidity ratio, the jointless gene (breaking the fruit without peduncle), and genetic resistance to the main pathogens, concentrated fruit maturation suitable for mechanized harvesting. The researches were completed for a total of 5 new varieties patented and registered in the Official Catalogue of the Crop Plants from Romania. Among these, the varieties Darsirius and Daria present ovoid, plump fruits, Kristinica and Florina 44 R have round fruits, and Florina 44 T has also round fruits that show an easy mucrone, transmitted by the beck gene. Regarding the earliness, the first place is occupied by the variety Kristinica, its fruits reaching the physiological maturity at 90 days after planting and the tardiest is the Darsirius variety, with fruits that reach the complete maturity at 130 days after planting.

Key words: Darsirius, breeding, Florina , germplasm, Kristinica.

INTRODUCTION

V.R.D.S. Buzău showed interest for tomatoes breeding since its establishment, in 1957. There were created the first Romanian varieties of tomatoes intended for both industry and fresh consumption. At the same time was obtained hybrid tomato seed for the first time, under the Bulgarian license, the famous hybrid 10 x Bizon, followed by Romanian Export II hybrid made from the crossing of line 24 x XIII.
Currently, the institution has a portfolio of 18 approved and patented varieties, registered in the Official Catalogue of Romanian Crop Plants. It was found that most Romanian varieties so far have a mixt destination, both for fresh consumption and industrialization. As a result of the increasing demands imposed by growers,

processors and consumers, starting with 1996, research undertaken intensively at V.R.D.S. Buzău in terms of tomatoes breeding had the aim of obtaining specialized strictly specialized varieties.
At the same time it was found that Romanian industry varieties offer is currently quite limited in comparison with market needs. In terms of per capita consumption tomatoes are the leading processed vegetables. (Gould et al., 2013). Tomatoes intended for processing must meet certain conditions, to have certain features that are implemented through the breeding process of improvement and genetic resource used.
For instance, certain phenological traits (early flowering and concentrated fruit set) were associated with a set of morphological traits (smaller canopies and low vegetative biomass),

along with gains in physiological traits (biomass N concentration and photosynthetic rates) in modern varieties (Felipe et al., 2014).

The breeding program aimed the fruit quality, especially in terms of their chemical composition. Processing of fresh tomato into paste had an overall positive effect on the contents in phenolic compounds, no effect on lycopene and a slight and high detrimental effect on β-carotene and ascorbic acid, respectively, (Chanforan et al., 2012).

This work presents recent varieties obtained for industry as results of research undertaken in 1996-2015 at V.R.D.S. Buzău

MATERIALS AND METHODS

The research started with the collection of germplasm, its evaluation and division by type (sp- self prunning, SP- half self prunnig and SP⁺- indeterminate) and depending on the degree of genetic stability (stable, advanced and segregant) (fig.1.).

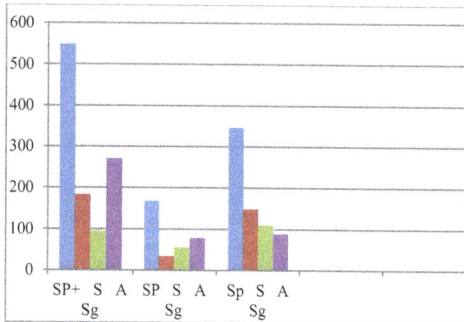

Figure 1. Tomatoes germplasm composition

SP+ (indeterminate)- 549 lines from which:
S (Stable) 183,
A (advanced) - 95
Sg (segregant) - 271
SP (half self-pruning lines) - 168 lines from which:
S (Stable) 34,
A (advanced) - 56
Sg (segregant) -78
Sp-(self-pruning lines)- 347 lines from which:
S (Stable) 148
A (advanced) - 110
Sg (segregant) -89

After germplasm evaluation, it was divided into two fields, the general collection field where were maintained the collected genotypes and work field composed from promoted genotypes in accordance with the breeding process objectives.

Priority had self-pruning genotypes genetically stabilised totalling 148.

Breeding applied methods were specific for tomatoes, especially individual repeated selection.

After the evaluation and tests carried out in the test fields over a long period of time, 5 genotypes showed stability and genetic superiority in terms of productivity and meeting specific industry characteristics and were approved under the following names: Daria (Măriuca), Florina R, Florina T, Darsirius and Kristinica.

As control variant in the experience has been used well-known variety Rio Grande.

The applied technology was specific to tomatoes specifying that the establishment of culture, for all the cultivars was done through direct sowing and planting stock. Sowing for seedling production has been carried out on the 10th of March and the planting was carried out on the 25th of April (Fig.2.).

Figure 2. Seedlings

For direct sowing variant, all cultivars were seeded on the 20th April.

Soil preparation was made in sSeptember through leveling, followed by fertilization. The work was followed by deep ploughing. In spring, the soil has been mobilized with disc harrows, followed by soil modeling. Maintenance work was specifically, irrigations (7-8) during the vegetation period, filling in gaps, manual and mechanical cultivation.

Planting has been carried out using the following scheme of crop establishment (fig.3.):

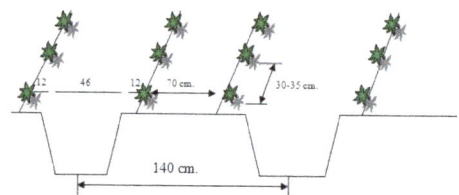

Figure 3. Crop establishment scheme on modeled soil

The same establishment scheme was used to direct sowing crop variant specifying that the norm of seed was 1 kg/ha, followed by sparing work, after the plants have reached the cross stage.

RESULTS AND DISCUSSIONS

The researches were completed with obtaining a rich and varied assortment of new Romanian varieties at this species. The new varieties have distinct genetic and phenotypic features that correspond to the proposed objectives for industrialization.

The main features of the newly created varieties are:

Florina R variety of tomato (fig. 4.) has determined growth intended for field crop. Special taste and aroma and also the dry matter percentage over 6.2% recommends it for industrialization. Immature fruit is green, with green shoulder (U gene) and at physiological maturity turns bright red. The fruit are firm, split and burning sun resistant and has a long shelf life after harvest (over 10 days). The fruit has commercial, attractive aspect with pleasant and balanced taste. The fruit can be jointed harvested because of the short stem, increasing the duration of storage after harvesting.

Plant vigour is average, presents a number of 60-80 leaves with leaflets of medium size.

In the inflorescence can be found 4-6 large, round fruit, medium weight of fruit is 220 g.

In cross section the fruit shows a pericarp thickness of 7-8 mm and 4 seminal lodges.

The fruit has a small number of seeds, between 60-80 that are well developed and visibly covered with yellow thin hairs.

Production potential = 50-60 t/ha.

The Florina T variety (fig. 5) is similar to the Florina R variety, having at the obtaining base the same origin but differentiates itself by the following characteristics: the fruit is round with easy mucron (Beck gene). At the same time the yield level is low but the shelf life is longer due to the higher firmness.

The Darsirius variety (fig. 6.) has ovular shape fruits with an average weight of over 80 g, immature fruits have uniform green colour (U.G. gene) and riped fruits are glossy dark red. The fruits are firm, with small abscission area without hard tissues into the fruit, jointless,

easily to be harvest and it is split resistant, reddish-burgundy colour in ripe. Due to the concentrate ripening and jointless fruits, the variety is suitable also for mechanized harvesting.

Fig. 4. Florina R variety

Fig. 5. Florina T variety

It has a high content of dry matter of more than 5.5 percent. It can be grown in organic farming, being created in our country concerning the specific climate conditions. Plant has determined growth with a height ranging between 50-60 cm with 4-6 vigorous shoots.

Figure 6. Darsirius variety

Variety is genetically endowed with genes of resistance to specific tomatoes diseases and Nematodes (Mi gene) and has a long shelf life after harvesting for over 10 days.

The Kristinica variety (fig. 7.) has a determined growth, registering an average height of 60 cm, small vigor and shallow foliage.

Figure 7. Kristinica variety

Thanks to this feature it can enlarge the surface density and shows an average number of 6 fruit per truss. Before maturity, the fruit presents green shoulders (U gene).

The fruit is round shaped and has an average weight of 120 g. Fruits are firm and are red colored. The shelf life of the fruit both on the plant and after ripened, after harvest is over 30 days, without yield depreciating. Yield production: 50-60 t/ha, 2, 5 kg/plant. Yield can increase significantly if it interferes with additional technological links.

Beeing the first obtained and patented variety of V.R.D.S. Buzau for industry, it was tested in the six main vegetable romanian institutions in comparative crop with Rio Grande. The obtained results are presented in table no.1.

Table 1. testing of potential yield production (t/ha) of tomato variety Kristinica in comparative crop, in six of the main romanian vegetable institutions

Variety	Locality						Average		STAS of total**	Early yield of total***
	Tc.	Ov.	Cl.	Cf.	Tu,	Tg.	t/ha	%	%	%
Rio Grande (control variant)	52.2	22.9	44.2	68.3	9.8	37.5	39.2	100	77.2	5.4
Kristinica	41.8	29.2	50.3	82,0	20.4	40.3	44	112.2	86.9	21.2

*Locality: Tc.=Tecuci; Ov.= Ovidiu;   DL 5%=8,8 t/ha
Cl.= Calarasi; Cf.= Calafat; Tu.= Turda   DL 1%=13,8 t/ha
Tg= Targoviste   DL 0,1%=23,5 t/ha
**Fruits with weight of over 33 g;
***Yield obtained until 31st of July- in the south; 10th of August- in the rest of area

Daria (Mǎriuca) variety (fig. 8) is determined growth, medium vigor, with an average height of 55 cm, a vigorous stem, from which 6-8 sprouts start. The leaves are medium sized, dark green, slightly embossed.

Figure.8. Daria (Mǎriuca) variety

The fruit is slightly ovular shape, immature fruit is uniform green (UG gene) and red riped. The fruits are firm, with a pericarp of 8-9 mm and a total of 3 seminal lodges where there are between 60-80 seeds. The shelf life of the fruit is good, for more than 30 days. Average production per plant is 2.5 kg but can increase

significantly if it interferes with additional technological links.

In terms of yields obtained at V.R.D.S. Buzău in 2010-2015 period, is noted that compared to Rio Grande control variant, the 5 varieties for industrialization have recorded higher yields (table.no. 2.).

Daria (Mǎriuca) registered in 2014, 108.9 t/ha, being the highest yield compared to Rio Grande who registered just 75.2 t/ha in the same year, and highest yield recorded by this variety.

The smaller productions were obtained in 2011 (21.5 tonnes/hectare at Florina R) but with the difference that the climatic conditions of that year were not favourable for the tomatoes.

On average, all 5 varieties are superior compared to the control variant with a maximum value of 17.8 percent registered by Daria (Mǎriuca).

In order to obtain an early yield the concentrated ripening fruit was aimed and Florina T recorded the highest percentage of total early yield, of 9.6%, while Florina R had 6.9 percent.

Table 2. Yield obtained at V.R.D.S. Buzău in 2010-2015

| Variety | Yield (t/ha) | | | | | | | | STAS* of total yield | Early yield of total** |
| | Year | | | | | | Average | | | |
	2010	2011	2012	2013	2014	2015	t/ha	%	%	%
Rio Grande	34.8	25.6	24.5	62.9	75.2	73.2	49.4	100	93.6	8.3
Florina T	33.4	25.5	36.0	54.2	71.8	63.1	47.3	95.7	94.3	9.6
Florina R	33.5	21.5	32.4	81.0	93.5	53.9	52.6	106.5	93.1	6.9
Daria	31.2	22.7	51.6	71.8	108.9	63.1	58.2	117.8	93.0	8.1
Darsirius	31.8	22.3	37.2	74.5	76.4	75.6	53.0	107.3	93.3	7.2
Kristinica	30.9	25.2	36.7	68.0	71.3	77.5	51.6	104.5	93.4	7.9

* fruits with weight of over 33 g
**Yield obtained until 31st of July

DL	5%	10.3	20.9 %
DL	1%	14.0	28.3 %
DL	0,1%	18.7	37.8 %

The industry varieties were biochemical analysed and it was found that in the case of dry matter content, due to the pericarp density, Darsirius variety ranked the first place with 5.8%; dry soluble matter remains at the rate of 5% as in the case of the variety Kristinica (table no. 3). In the case of varieties of Florina T, R and r. Daria (Măriuca) an equal value of 4.5% was recorded. Also, Kristinica has registered a value of 0.43% acidity, at the opposite side being Darsirius with 0.35%. The highest content in sugar was recorded by Darsirius while Daria (Măriuca) has registered a rate of 2.44 percent.

Regarding the sugar : acidity ratio, Darsirius was first valued with 8.97%, followed by Florina R with 16%. The highest content in ascorbic acid was measured at Florina T followed by 11.97% at Darsirius with 9.58%. The highest content of lycopene 9.08% was recorded by Florina R.

Table 3. Biochemical analysis of tomatoes for industrialization

| Parametre | Variety | | | | | |
	Daria (Măriuca)	Kristinica	Florina T	Florina R	Darsirius	Average values*
d.m.c. %	5.18	5.26	5.35	5.30	5.8	5.98±0.83 (cv=14%)
s.m.c. %	4.5	5.00	4.5	4.50	5.00	4.75±0.35 (cv=7%)
Acidity (Citric acid), %	0.41	0.43	0.39	0.37	0.35	0.38±0.03 (cv=8%)
Sugar total,%	2.44	2.97	2.88	3.02	3.14	3.37±0.77 (cv=21%)
Ratio sugar: acidity	5.95	6.91	7.38	8.16	8.97	10.76±0.09 (cv=1%)
Ascorbic acid, mg/100 g^{-1}	8	7.02	11.97	8.30	9.58	9.67±0.65 (cv=7%)
Lycopene, mg/100 g^{-1}	6.5	5.00	8.18	9.08	6.00	8.00±1.5 (cv=19%)
Average weight,g± a.s.	93.5±22.1 (cv=24%)	108.9±13.5 (cv=12%)	139.9±29.1 (cv=21%)	167.9±26.0 (cv=15%)	86.9±13.3 (cv=15%)	92±10 (cv=11%)

* Viorica and Vipon varieties (RDIVFG Vidra)
d.m.c. – dry matter content
s.m.c. – soluble matter content

CONCLUSIONS

The researches were completed so far with the achievement of a germplasm resource at this species consisting of 1064 genotypes, both evaluated and computerized stored from which in the future will be able to obtain new varieties. 5 new varieties of tomatoes for industrialization have been obtained and approved: Kristinica, Darsirius, Florina T and R and Daria (Măriuca), genotypes with distinct features that enrich the current industry tomatoes assortment.

All five varieties behaved positively to the both crop systems direct sowing and planting stock specifying that the plants were more vigorous in the direct sowing crop, but with 15-20 days later yield. Recorded yields and physical and chemical properties of fruits demonstrates that the objectives of the proposed research breeding program were reached.

REFERENCES

Chanforan C.., 2012. The Impact of industrial processing on health-beneficial tomato microconstituents, Food chemistry vol. 134, issue 4, 1786-1795.

Felipe H. Barrios-Masias, Louise E. Jackson, 2014. California processing tomatoes: Morphological, physiological and phenological traits associated with crop improvement during the last 80 years. European Journal of Agronomy, vol 53, 45-55.

Gould W.A., 2013, Tomato production, processing and technology. 3rd Edition Elsevier, USA

2

THE INFLUENCE OF CULTURE TECHNOLOGY UPON THE PHYSICAL QUALITY OF SOME EARLY TOMATOES VARIETIES

Constanța ALEXE[1], Marian VINTILĂ[1], Simona POPESCU[1], Gh. LĂMUREANU[2], Lenuța CHIRA[3]

[1]Research and Development Institute for Processing and Marketing of the Horticultural Products – HORTING Bucharest, No. 1A, Intrarea Binelui Street, District 4, 042159, Bucharest, Romania, E-mail: ihorting@yahoo.com

[2]Research Station for Fruit Growing (R.S.F.G) Constanta, No.1, Pepinierei Street, 907300, Commune Valu lui Traian, Romania E-mail: scpp_constanta@hotmail.com

[3]University of Agronomic Sciences and Veterinary Medicine of Bucharest, No. 59, Marasti Blvd, District 1, 011464, Bucharest, Romania, E-mail: lenutachira@yahoo.com

Corresponding author email: constanta_alexe@yahoo.com

Abstract

The quality of the Romanian vegetable production is currently of a great importance as far as alimentation, horticultural economy and commerce with such perishable products because that determines competition on both internal and external market and, implicitly, the maintaining of the market for Romanian products in the context of an open, competitive market. Our researches aimed to establish the most appropriate culture technological sequences for three varieties of early tomatoes ('Isalnita 29', 'Isalnita 50', 'Buzau 47') in order to obtain high quality fruit with suitable physical qualitative indicators. All tomatoes varieties that were tested benefited in culture for three different density variants (25,000 plants/ha, 40,000 plants/ha, 55,000 plants/ha) and two levels of fertilization (c1 = N:200 kg/ha; P2O5:100 kg/ha, K2O:100 kg/ha, c2 = N:300 kg/ha; P2O5:200 kg/ha; K2O:100 kg/ha). Immediately after harvesting, certain physical determinations were carried out concerning the main physical qualitative indicators of the fruit: average weight, thickness of the pericarp, specific weight and texture firmness. Results show that the physical qualitative indicators vary depending on variety, planting density and lightly on fertilizer dose of culture. Between the three varieties that were studied, the variety 'Buzau 47' is distinguished through the largest fruits (average weight=97.75 g), high specific weight (0.9726 g/cm3) and the thickness of the pericarp (6.66 mm). At the same time, the variety 'Buzau 47' has the fruits with the lowest firmness (145.87 PU), this indicator having values inversely proportional to the size of fruits. Regarding the planting density, this influences, according to the physical qualitative indicator, in a different way. As the density is lower, the average weight of the fruit has higher values. Between tested fertilization variants, at a level of nutrition below the limits of 300 kg/ha N, 200 kg/ha P2O5 and 100 kg/ha K2O, there are no essential differences in the values of the main physical qualitative indicators, beside the average weight of the fruits, which increases from 69.97 g in the case of fertilization variant c1, to 82.92 g in the case of fertilization variant c2.

Key words: average and specific weight, firmness, level of nutrition, planting density.

INTRODUCTION

Tomatoes are one of the most important vegetable species in our country, due to the fact that they can be consumed both fresh and processed in different ways (Stan et al., 2003).

Tomatoes are healthy and contain very few calories. They have a significant content of vitamin C, minerals (e.g.: potassium) and important micro-nutrients.

Supplying market with fresh tomatoes obtained in open field is possible beginning with the second half of June by performing early cultures. This type of culture, which occupies a significant share in our country, is practiced in the areas with more favourable climatic conditions for tomatoes, such as: Western Plain, Danube Plain, a part of Dobrogea.

Researcher Vînătoru (2006) affirmed that the Romanian tomato is tasteful, aromatic and beneficial for health, being cultivated on natural soil, not forced with chemical substances.

The cultivating method on unconventional substrata is still relatively new in Romania (Makobo and Du Plooy, 2008), so that the classical method, on soil, still occupies the largest part of the surface in our country destined for tomato cultures (Ciofu et al., 2004). The aim of the applied different culture technologies has to not only be the obtaining of large productions, but also to ensure a high quality, which means that the technological links have to take into account the destination of the production. From this point of view the fertilization system and regime have significant effects (Neata, 2002; Cioroianu et al., 2010; Anton, 2011; Cioroianu et al., 2011). Also at the creation of new varieties and hybrids should be taken into consideration the fact that they respond differently to both environmental conditions and technological links applied to the culture (Draghici and Pele, 2012).

In appreciation of the quality and nutritive-alimentary value of the fruits, is taken into consideration the physical and sensory characteristics (size, shape, colour, specific weight, texture firmness, flavour, taste etc), technological characteristics (storage capacity, transport and handling resistance, presence of diseases or pests attack, remanence of pesticides) and the biochemical properties: water content, dry matter, carbohydrates, acids, cellulose, vitamins, pigments, mineral salts (Salunkhe et. Kadam , 1998; Alexe et al., 2013).

This paper presents some aspects regarding the influence of variety, planting density and fertilization of early tomato culture upon the certain physical qualitative indicators of the fruits.

MATERIALS AND METHODS

The researches were conducted during period 2013-2014, using Romanian varieties of early tomatoes, obtained in a vegetable farm located in an area of the Romanian seaside.

The trial was organized as a trifactorial experience, with following experimental factors:

A – planting density (plants/ha)	B – variety	C – ferilization level (kg/ha)
a1 – 25,000	b1- Isalnita 29	c1 – N:200; P_2O_5:100; K_2O:100
a2 – 40,000	b2 - Isalnita 50	c2 – N:300; P_2O_5:200; K_2O:100
a3 – 55,000	b3 - Buzau 47	-

The observations and determinations regarding the main physical qualitative indicators (average weight, thickness of the pericarp, specific weight, texture firmness) were made at Research and Development Institute for Processing and Marketing of the Horticultural Products - Horting Bucharest and at University of Agricultural Sciences and Veterinary Medicine Bucharest. The determination of the fruit firmness was performed by means of a mass penetrometer OFD, the measurement being in penetrometer units (1PU = 0.1 mm) of the depth of the conical needle penetration (length = 24mm, diameter at base = 4 mm) in the pulp. Measurements were performed on a total of 25 fruit/variant, each fruit being penetrated in four points in the equatorial zone.

RESULTS AND DISCUSSIONS

The average weight of the fruit is a characteristic indicator for every variety. Between 3 varieties of early tomatoes that were studied, taking into account the average of variants, Isalnita 29 variety has the smallest fruits, with the average weight of 51.78 g, varying, depending on the distance of planting and fertilization variant, between 51.0 g and 53.3 g (Table 1).

Variety Isalnita 50, with the value of average weight of fruits of 79.80 g, has much larger variation limits depending on the variant of culture (67.0-93.1 g).

The largest fruits are found at variety Buzau 47, with the average weight of 97.75 g and variation limits between 76.0 g and 108.3 g.

Specific weight is high at all 3 varieties, the average value being between 0.9256 g/cm^3 at variety Isalnita 29 and 0.9726 g/cm^3 at the variety Buzau 47. Higher differences between variants were observed at variety Buzau 47, with limits from 0.8876 g/cm^3 to 1.1028 g/cm3, while the variety Isalnita 29 presented more constant values around the average (0.8903-0.9712 g/cm^3).

Table 1. The influence of variety upon physical qualitative indicators of early tomatoes

Variety	Variant	Average weight (g)	Specific weight (g/cm³)	Thickness of pericarp (mm)	Firmness (PU)
b1	a1c1	52.1	0.9044	5.5	91.8
	a2c1	52.0	0.8916	5.5	108.3
	a3c1	52.0	0.8903	5.3	106.3
	a1c2	52.3	0.9941	5.5	102.5
	a2c2	51.0	0.9712	5.7	107.5
	a3c2	51.3	0.9021	5.8	92.0
	average	51.78	0.9256	5.55	101.40
b2	a1c1	80.9	0.9614	6.6	131.7
	a2c1	81.8	0.9404	6.5	140.3
	a3c1	67.0	0.9645	6.5	130.5
	a1c2	93.1	0.9761	6.6	130.4
	a2c2	88.1	1.0091	6.6	135.4
	a3c2	67.9	1.0041	6.4	103.5
	average	79.80	0.9759	6.53	128.63
b3	a1c1	108.3	0.8876	6.6	151.1
	a2c1	83.1	0.9334	6.6	147.5
	a3c1	76.0	0.9019	6.5	130.5
	a1c2	114.1	0.9881	6.8	148.3
	a2c2	108.3	1.0221	6.8	150.5
	a3c2	101.2	1.1028	6.7	147.3
	average	97.75	0.9726	6.66	145.87

The thickness of pericarp is, as well, a character of variety, being lower at Isalnita 29 variety, of 5.55 mm (which presents also the smallest weight of the fruits) and very close at the other 2 varieties (6.53 mm at variety Isalnita 50, respectively 6.66 mm at variety Buzau 47). There were not registered differentiated values within the variants.

Firmness of the pulp presented values inversely proportional to the size of fruit, being the lowest at variety Buzau 47 (145.87 PU), the highest at variety Isalnita 29 (101.40 PU) and with intermediary values at variety Isalnita 50 (128.63 PU).

The results presented in Table 2 show that the planting density is also influencing some physical qualitative indicators of fruits. The average weight is higher at a lower planting density.

At a planting density of 25,000 plants/ha, average weight had the average value of 82.31g, while at a planting density of 55,000 plants/ha, this was only 76.16 g. However from table 2 results that the indicator average weight is influenced by the culture density only at varieties Isalnita 50 and Buzau 47, whose fruits are smaller as the density increases.

Table 2. The influence of planting density upon physical qualitative indicators of early tomatoes

Planting density	Variant	Average weight (g)	Specific weight (g/cm³)	Thickness of pericarp (mm)	Firmness (PU)
a1	b1c1	52.1	0.9044	5.5	91.8
	b2c1	80.9	0.9614	6.6	131.7
	b3c1	98.3	0.8876	6.6	151.1
	b1c2	52.3	0.9941	5.5	102.5
	b2c2	93.1	0.9761	6.6	130.4
	b3c2	117.2	0.9881	6.8	148.3
	average	**82.31**	**0.9519**	**6.27**	**125.96**
a2	b1c1	52.0	0.8916	5.4	108.3
	b2c1	81.8	0.9404	6.5	140.3
	b3c1	93.1	0.9334	6.6	147.5
	b1c2	52.3	0.9712	5.5	102.5
	b2c2	88.1	1.0091	6.6	135.4
	b3c2	114.1	1.0221	6.8	150.5
	average	**80.23**	**0.9613**	**6.23**	**130.75**
a3	b1c1	51.0	0.8903	5.3	116.3
	b2c1	76.0	0.9642	6.5	148.5
	b3c1	94.5	0.9019	6,7	155.3
	b1c2	51.3	0.9021	5.8	112.0
	b2c2	75.9	1.0041	6.4	143.5
	b3c2	108.3	1.1028	6.7	157.3
	average	**76.16**	**0.9609**	**6.23**	**138.81**

The average weight of fruits from variety Isalnita 29 presented constant values, regardless of the culture density. This variety allows therefore higher culture densities, without being affected the uniformity of production, while, at the other 2 varieties, the density of 55,000 plants/ha may lead to unevenness of average weight and implicitly of production.

The firmness of fruits, which is a variety distinctiveness, presents a great importance for a superior valorification, for which it is necessary to be given attention in the application of technological links of culture. This indicator is influenced by the planting density. The fruit firmness decreased from the value of 125.96 PU at variant a1 to 130.75 PU and respectively to 138.81 PU in case of variants a2 and a3 respectively. The softest fruits at harvest are met therefore at the culture density of 55,000 plants/ha, at variety Buzau 47. The thickness of pericarp is also a physical indicator of a great importance, which, in the case of mechanical conditioning, should be taken into consideration. This indicator did not presented significant modifications along with the increasing of planting density.

Specific weight and thickness of the pericarp was less influenced by planting density. However, variant a1 presented a slight decreased value of specific weight (0.9519 g/cm³) comparatively with variant a2 (0.9613 g/cm³) and variant a3 (0.9609 g/cm³).

The influence of fertilization levels upon physical indicators is represented in Table 3.

The average weight of fruits is influenced by the level of fertilization doses, that increasing, from 69.97 g in the case of fertilization variant c1, to 82.92 g in the case of fertilization variant c2, considering the average of variants planting density x variety. It is observed repeatedly the stability of Isalnita 29 variety, to whom the modifications from c1 to c2 are insignificant.

Table 3. The influence of fertilization level upon physical qualitative indicators of early tomatoes

Fertili-zation level	Variant	Average weight (g)	Specific weight (g/cm³)	Thickness of pericarp (mm)	Firmness (PU)
c1	a1b1	52.1	0.9044	5.4	91.8
	a1b2	71.9	0.9614	6.6	131.7
	a1b3	98.3	0.8876	6.8	151.1
	a2b1	52.0	0.8916	5.4	108.3
	a2b2	70.8	0.9404	6.6	140,3
	a2b3	84.1	0.9334	6.6	147.5
	a3b1	51.0	0.8903	5.3	106.3
	a3b2	66.0	0.9642	6.5	130.5
	a3b3	83.5	0.9019	6.6	145.3
	average	69.97	0.9194	6.18	128.09
c2	a1b1	52.3	0.9941	5.5	107.5
	a1b2	93.1	0.9761	6.6	130.4
	a1b3	117.2	0.9881	6.9	160.5
	a2b1	51.3	0.9712	5.7	109.5
	a2b2	88.1	1.0019	6.6	135.4
	a2b3	114.1	1.0221	6.7	148.3
	a3b1	51.0	0.9021	5.8	115.0
	a3b2	75.9	1.0041	6.4	143.5
	a3b3	103.3	1.1028	6.7	157.3
	average	82.92	0.9958	6.32	131.93

Concerning the specific weight, thickness of the pericarp and firmness of the pulp, there are no essential differences between the 2 levels of fertilization. It is observed however a slight

increase of the average values of specific weight and pericarp thickness and a small decrease of the fruits firmness in the case of variant c2 beside c1.

Results concerning the interaction of factors A x B x C show that the fertilization acts independently of variety and planting density, but in both variants, variant c1 and variant c2, the higher value of average weight (98.3 g respectively 117.2 g), thickness of the pericarp (6.8 mm, respectively 6.9 mm), and texture firmness (151.3 PU, respectively 160.5 PU) was recorded at variant a1b3 (a1: planting density=25,000 pl/ha; b3: variety Buzau 47), proving the influence of variety and planting density (Figure 1).

Fig. 1. The influence of interaction of the factors variety (Buzau 47), planting density (25,000 pl/ha) and fertilization level upon some physical qualitative indicators of the early tomatoes

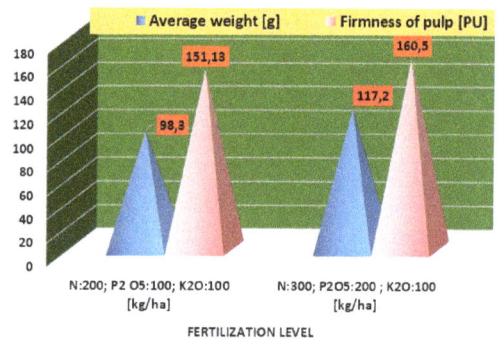

CONCLUSIONS

The main physical qualitative indicators (average weight, thickness of the pericarp, specific weight, texture firmness) varies depending on variety and culture technology conditions.

Between the three varieties that were studied, the variety Buzau 47 is distinguished through the largest fruits (average weight=97.75 g), high specific weight (0.9726 g/cm³) and the thickness of the pericarp (6.66 mm). At the same time, the variety Buzau 47 has the fruits with the lowest firmness (145.87 PU), this indicator having values inversely proportional to the size of fruits.

Regarding the planting density, this influences, according to the physical qualitative indicator, in a different way. As the density is lower, the average weight of the fruit has higher values. From this point of view, the tomatoes that came from culture with plantig density of 25, 000 plants/hectars recorded the best results. However the average weight of fruits from variety Isalnita 29 presented constant values, regardless of culture density. This variety allows therefore higher culture densities, without being affected the uniformity of production, while, at the other 2 varieties the density of 55,000 plants/ha may lead to unevenness of average weight and implicitly of production. Along with the increasing of density, the firmness decreases, the softest fruits at harvest are meeting at the culture density of 55,000 plants/ha.

In the case of different fertilization level, at a nutritional level below the limits of 300 kg/ha N, 200 kg/ha P_2O_5 and 100 kg/ha K_2O, there are no essential differences in the values of the main physical qualitative indicators, beside the average weight of the fruits. This increases, from 69.97 g in the case of fertilization variant c1, to 82.92 g in the case of fertilization variant c2, having into consideration the average of the variants planting density x variety. It is observed repeatedly the stability of Isalnita 29 variety, to whom the modifications from variant c1 to variant c2 are insignificant.

REFERENCES

Alexe Constanta, Lamureanu Gh., Chira Lenuta, Pricop Simona, 2013. The influence of culture technology upon the temporary storage capacity of tomatoes. Journal of Horticulture, Forestry and Biotechnology, vol 17 (3) - Banat University of Agricultural Sciences and Veterinary Medicine Timosoara: 91-96

Anton Iulia, Dorneanu A., Bireescu Geanina, Sîrbu Carmen, Stroe Venera, Grigore Adriana, 2011. Foliar fertilization effect on production and metabolism of tomato plants. Research Journal of Agricultural Science, 43 (3): 124 131

Ciofu Ruxandra, Draghici Elena, Dobrin Elena, 2004. Legumicultura speciala. Indrumator de lucrari practice. Editura Elisavaros, Bucuresti, 53-56

Cioroianu T., Sîrbu Carmen, Dumitrascu Monica, Stefanescu S., 2010. Fertilizanti organo-minerali cu utilizare in agricultura durabila. Simpozionul stiintific anual cu participare internationala, "Horticultura - stiinta, calitate, diversitate si armonie", Iasi, Lucrari stiintifice USAMV Iasi, seria Horticultura, Vol. 52, pp 304-310

Cioroianu T., Pohrib C., Sirbu Carmen, Grigore Adriana, Oprica Ioana, Mihalache Daniela, Anton Iulia, 2011. Assessment of quality tomatoes grown in solar by applying organic and mineral fertilization – Amanda hybrid, Book of abstracts Sesiunea Omagială - Agrochimia - Prezent şi viitor a Filialei Naţionale Romane CIEC, pp 72-80

Draghici Elena, Pele Maria, 2012. Evaluation some new hyrbid for cultivation in convencional system in springclimatic conditions of Romania, International Journal of Agriculture Science, vol.4, p.79-94

Makobo, M. M., Du Plooy, 2008. Comparative performances of tomato grown on soil vs in-soil production systems. International Symposium on Soiless Culture and Hyroponics, Peru, Lima.

Neata Gabriela, 2002. Agro-chemistry and soil biology. Printech Publishing House, Bucharest.

Salunkhe, D.K., Kadam S.S., 1998. Handbook of Vegetable Science and Technology: Production, Compostion, Storage and Processing. CRC Press: 171-203

Stan N., Munteanu N., Stan T., 2003. Vegetable growing, Vol. III, Ion Ionescu de la Brad Publishing House, Iasi

Vanatoru C., 2006. Crearea de hibrizi F1 de tomate timpurii cu plasticitate ecologica si calitate superioara. Teza doctorat.

3

THE BEHAVIOR OF SWEET POTATO *(IPOMOEA BATATAS)* IN TERMS PSAMOSOILS IN SOUTHERN ROMANIA

Aurelia DIACONU[1], Cho EUN-GI[2], Reta DRĂGICI[1], Mihaela CROITORU[1], Marieta PLOAE[1], Iulian DRĂGHICI[1], Milica DIMA[1]

[1]Research - Development Center for Agricultural Plants on Sands, Dabuleni, Dolj County, Romania
[2] Kyungpook National University in South Korea

Corresponding author email: aureliadiaconu@yahoo.com

Abstract

Variability of climate, especially lack of rain and low fertility psamosoils determines that surfaces quite stretched in most crop yields to be much reduced. In this context, the choice assortment of plants and varieties with high adaptability to the harsh conditions of climate and soil is a necessity for obtaining high yields, stable and reliable, that provide increased energy needs of the population food. Research conducted during 2013-2015 at the Research - Development Center for Agricultural Plants on Sands, Dabuleni, highlights a favorable microclimate for plant growth and development of sweet potato (Ipomoea batatas). The values recorded physiological indices sweet potato grown in climatic conditions in the sandy soils of southern Oltenia showed that it easily adapts to the conditions of excess heat here is a plant heat-loving and light. Productions made from sweet potato varieties Pumpkin (KSP1) and Chestnut (KSC 1), sandy soil conditions studied ranged 17428kg / ha and 35467 kg / ha depending on the crop and cultivated variety. We showed correlations between tuber production and the amount of degrees of temperature recorded in air (r = 0.904; r = 0.992). Also between rainfall and production carried out by two varieties of sweet potato is negative correlations (r = -0.642; r = - 0.848). These correlations highlight the specificity of the plant to dry climate. Nutritional quality presented values differentiated to the two varieties studied, according to the year of culture.

Key words: sweet potato, tolerant, drought, physiology, quality

INTRODUCTION

Sweet potato (*Ipomoea batatas*) belongs to the family *Convolvulaceae* and is native to Central America and the north-west of South America. Globally, is among the food crops the most important in the world after wheat, rice, corn, potato Irish and barley, being adaptable to climate tropical and subtropical zones, drought-tolerant and grows in conditions of fertility and soil pH decreased (Kareem I., 2013). It is a plant well adapted to tropical and subtropical climates, but can grow successfully in a wide range of climatic conditions in the cold season average to not more than 5 months. Research by James A. Duke 1983 plant underlines the sensitivity at low temperatures, frost tolerating plant (www.ncsweetpotatoes.com). It grows best at an average temperature of 24°C, with abundant sunshine and warm nights. Annual rainfall of 750-1000 mm are considered most appropriate, with a minimum of 500 mm in the growing season. Culture is sensitive to drought tuber initiation stage, 50-60 days after planting and is not tolerant to water stagnation, it can cause tuber rots and reduce root growth due to poor aeration. Abundant in nutrients and fiber (of which 40% soluble fibre that helps lower the blood sugar and cholesterol), sweet potato is the ideal food for diabetics, children and pregnant women (Betty J. Burri, 2011 Mihaela Cioloca et al., 2013). Through its qualities, a sweet potato variety with yellow and orange pulp is a valuable source of vitamin A and vitamin B6. Also, sweet potato provides a significant amount of vitamin C and vitamin D, essential in the formation of bones and teeth, for good digestion, wound healing and immune system. Sweet potato also contains iron, which helps metabolize protein, and magnesium, a mineral stress. Concerns for sweet potato cultivation in Romania took place in USAMV Bucharest (Ciofu R., 2005, Musat C., 2010) and were resulting in the creation of two varieties (Crux V., 1991, 1997). The variety of sweet potato varieties grown in Romania is quite limited and they are cultivated more experimental. Orange and yellow varieties have a high content of beta-carotene, the precursor of vitamin A (Kareem I., 2013). Therefore encourages the cultivation of these species in places like Africa where vitamin deficiency is causing severe health problems (Wariboko C. et al., 2014, Ladokum OA, et al., 2007). All species of sweet potatoes are rich in antioxidants. Sweet potato, although it has a sweetish taste, the presence of complex carbohydrates, which help regulate carbohydrate and reduced insulin resistance, is beneficial for people with diabetes. Based on these considerations nutrients, Center for Science in the Public Interest of America awarded the highest score compared to other vegetables. They also have a low glycemic index, which means that hunger will appear later. Research - Development Center for Agricultural Plants on Sands, Dabuleni began in 2013 a collaboration with the Institute for Agricultural Technology and Science Kyungpook National University, based in Daegu, represented by Prof. dr. Jong - and Dr. Sang KIM Gi Cho Eun-, South Korea, under "Technical and Scientific Cooperation Memorandum between Kyungpook National University ASAS Bucharest and the Institute for Agricultural Science and Technology. Within this collaboration was initiated research on the behavior of two varieties of sweet potato (Ipomoea batatas): Pumpkin (KSP1) and Chestnut (KSC1) in terms of climate and soil from RDCAPS Dabuleni, Dolj County.

MATERIALS AND METHODS

Research at the culture of sweet potato (Ipomoea batatas), were conducted in 2013-2015 at RDCAPS Dabuleni on sandy soil with low natural fertility (0.42 to 0.82% humus) and pH (H_2O) = 5.9 to 6.9 in the Bilateral Cooperation Protocol between Kyungpook National University (KNU) in South Korea and the Academy of Agricultural and Forestry Sciences "Gheorghe Ionescu sisesti" Bucharest. It studied the behavior of two varieties Korean: Pumpkin (KSP1) and Chestnut (KSC1) in terms of climate and soil in the sandy soils of southern Oltenia. Seedling product to the protected system type solar greenhouses as follows:
-between 20 to 25 March were planted sweet potatoes from seed in the greenhouse (Figure 1)

Figure 1. Planting sweet potato tubers in greenhouse

-between March and April are maintained in the greenhouse through irrigation and ventilation
After cutting the shoots, fertilize the solar N50 and watered with water necessary to obtain another generation of shoots (Figure 2)

Figure 2. Cutting the shoots in solar

-from May 5 to 10 they were cut shoots produced in solar and were planted in the field billon covered with polyethylene mulch. (Figure 3).

Figure 3. Planting the sweet potao shoots in field

In field was fertilized with $N_{80}P_{80}K_{80}$, to prepare the ground and in the vegetation was fertilized with N_{70}.Sweet potato tubers grown in two varieties were received from the Republic of Korea in accordance with the cooperation agreement. During the growing season, in root tubers stage was determined the photosynthesis and transpiration rate of plant to leaf level with + Portable Photosynthesis System LCpro device in three times of the day. At harvest the production of tubers has been determined and the quality of production as follows: total dry substance and water (%) - gravimetric method; simple soluble carbohydrates (%) - Soxleth Reagent method; the starch (%) - colorimetric method; C vitamin (mg/100 g f.s.) - iodometric method.

RESULTS AND DISCUSSIONS

Climatic conditions during the period 2013-2015 reveals an increase in atmospheric drought, revealing the sweet potato vegetation period mean monthly temperature higher by approx. 0.15 to 1.45°C, compared to the annual average (Table 1). Although the average amount of rainfall recorded during the study period was 372.3 mm, with 102.92 mm above the annual average, they were unevenly distributed and water to meet the needs of the

work was drip irrigation necessary to ensure the ceiling of 80% of active moisture range. In terms of ensuring the thermal requirements for growing sweet potato, 2015 it was the most favorable, followed by 2013 and then 2014. The average air temperature during the growing season (May to September) was between 20.48 -21.78 °C and rainfall were within 269.2 to 516.9 mm. 2014 was unfavorable for sweet potato and due to heavy rainfall that occurred in September, which coincided with the maturation period tubers, preventing dry substance accumulation and weight gain formed tubers per plant. Also, the average temperature in 2014 was lower by 1.02- 1.30 °C, the values recorded in the other two years, which led to unevenness maturing tubers and finally to production results low, sweet potato is a heat-loving plant. To mature, sweet potato varieties needed about 3133.44 to 3332.34 °C, built up during the growing season.

Table 1. Climate characterization of sweet potato vegetation period (2013-2015 Dabuleni)

Month	Climatic element	2013 Year	2014 Year	2015 Year	Multiannual average
May	Air temperature average ^{0}C	20	16.6	19.2	16.8
	Rainfall average (mm)	61	117.4	52.4	61.6
	Relative humidity %	71	76.5	73	
June	Air temperature average ^{0}C	22.1	20.7	20.5	21.6
	Rainfall average (mm)	105.2	92	134.2	68.5
	Relative humidity %	76.8	76.9	73.8	
July	Air temperature average ^{0}C	23.5	23.1	24.9	23.1
	Rainfall average (mm)	36.2	125.6	11.0	54.2
	Relative humidity %	73.7	77.9	62.9	
August	Air temperature average ^{0}C	24.1	23.7	24.3	22.4
	Rainfall average (mm)	30.8	16	48.4	37.7
	Relative humidity %	68	72	68.2	
September	Air temperature average ^{0}C	17.8	18.3	20.1	17.75
	Rainfall average (mm)	36	165.9	84.8	47.38
	Relative humidity %	72.5	82,3	77.3	
Air temperature average during the growing sweet potato 01.05– 30.09 (^{0}C)		21.5	20.48	21.78	20.33
Rainfall average during the growing sweet potato 01.05– 30.09 (mm)		269.2	516.9	330.8	269.38
Relative humidity average during the growing sweet potato 01.05– 30.09 (%)		72.4	77.12	71,04	-
$\sum {}^0$C in air during the growing sweet potato 01.05– 30.09		3292.6	3133.44	3332.34	3110.49
Rainfall (mm) during the growing sweet potato 01.05– 30.09		269.2	516.9	330.8	269.38
Production of sweet potato tubers (kg/ha)	Sweet potato variety / Year	2013	2014	2015	Average
	Pumpkin (KSP1)	23864	17428	33300	24864
	Chestnut (KSC1)	30176	18857	35467	28166

Results on physiological reactions to sweet potato depending on climatic factors:

During the growing season, in root tubers stage was determined the photosynthesis and transpiration rate of plant to leaf level with + Portable Photosynthesis System LCpro device in three times of the day (Figure 4). Figure 5 shows the diurnal variation of photosynthesis (micromoles CO_2 / m^2 / s) - determined the two varieties of sweet potatoes planted Pumpkin (KSP1) and Chestnut (KSC1) depending on the climatic conditions in the crop year 2014 -2015.

In 2014 the variety Pumpkin (KSP1) when determining the 15 o'clock when the daytime temperature reaches the highest values (38 -40 °C), photosynthesis rate recorded 12,555 micromoles CO_2 / m^2 / s and the variety Chestnut (KSC1) at the same time of day photosynthesis rate was 18,975 micromoles CO_2 / m^2 / s. 2015 variety pumpkin (KSP1) when determining the 15 o'clock rate of photosynthesis recorded 21 715 micromoles CO_2 / m^2 / s and the variety Chestnut (KSC1) at the same time of day rate of photosynthesis was 27 175 micromoles CO_2 / m^2 / s. Every year the production was positively correlated with the amount of diurnal variation of photosynthesis.

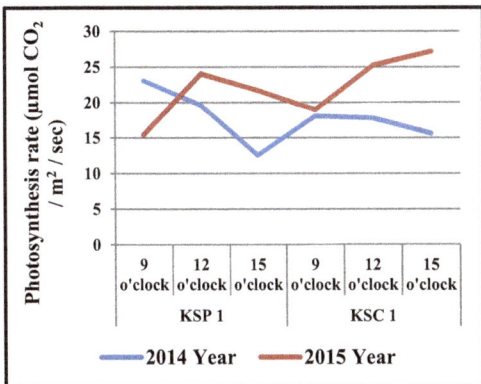

Figure 4. Diurnal variation of photosynthesis sweet potato plant, depending on the variety and the crop year

Figure no. 6 shows the diurnal variation of photosynthesis in the two sweet potato varieties in the 3 times of day in 2014 and 2015. In 2014 when the amount of degrees the temperature was lower than the annual average and exceeded the annual average

rainfall amount, variety Pumpkin (KSP1) recorded higher values of photosynthesis diurnal variation in measurements performed at 9:12 and at 15, the variety Chestnut (KSC1) achieved the highest value of diurnal variation of photosynthesis. Climatic conditions during the vegetation period of 2015 influenced the different speed and pacing of physiological processes to sweet potato grown in the sands. In July, air temperature varied between 26.7 to 41.5 °C determinations when air humidity fell below 25% and photosynthetic active radiation varies between 1200 to 1700 mol / m2 / s. In 2015 when the amount of degrees the temperature was higher than the annual average, and the amount of rainfall was lower than the annual average, the variety Chestnut (KSC1) recorded higher values of diurnal variation of photosynthesis in tests carried out in all 3 points of the day. This indicates that the variety Chestnut (KSC1) has the capacity to behave better in drought conditions than the variety Pumpkin (KSP1).

Diurnal variation of the sweet potato leaf transpiration depending on the variety and the crop

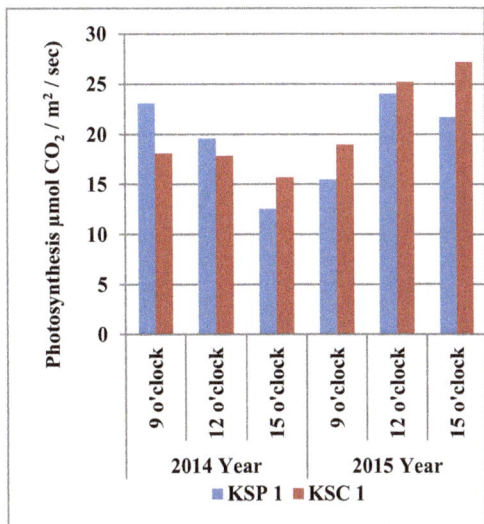

Figure 5. Diurnal variation of photosynthesis sweet potato plant, depending on the crop year and variety

Figure no. 7 presents diurnal variation of leaf transpiration (mmol H_2O / m^2 / s) - determined the two sweet potato varieties planted, Pumpkin (KSP1) and Chesnut (KSC1) depending on the climatic conditions in culture in 2014 -2015. In 2014 the variety Pumpkin (KSP1), diurnal variation of leaf transpiration (mmol H_2O / m^2 / s) when determining from 9 was 3.32 mmol H_2O / m^2 / s at 12 was 3.44 mmol H_2O / m^2 / s and at 15, when daytime temperatures reach the highest values, diurnal variation of leaf transpiration recorded 2.3 mmol H_2O / m2 / s. Variety Chestnut (KSC1) diurnal variation of transpiration leaf (mmol H_2O / m^2 / s) when determining from 9 was 2.795 mmol H2O / m2 / s at 12 was 2.975 mmol H_2O / m^2 / s and at 15 pm, when daytime temperatures reach the highest values, diurnal variation of leaf transpiration of 4.46 mmol H_2O / m^2 / s.

In 2015 Year, Pumpkin variety (KSP1), diurnal variation of leaf transpiration (mmol H_2O / m^2 / s) when determining from 9 was 1.97 mmol H_2O / m^2 / s at 12 was 7.01 mmol H_2O / m^2 / s, and at 15, when daytime temperatures reach the highest values, diurnal variation of transpiration foliar registered 9.705 mmol H_2O / m^2 / s the variety Chestnut (KSC1) diurnal variation of transpiration leaf (mmol H_2O / m^2 / s) when determining from 9 was 2.58 mmol H_2O / m^2 / s at 12 was 7.11 mmol H_2O / m^2 / s and at 15, when daytime temperatures reach the highest values, diurnal variation of leaf transpiration recorded 9.93 (mmol H_2O / m^2 / s).

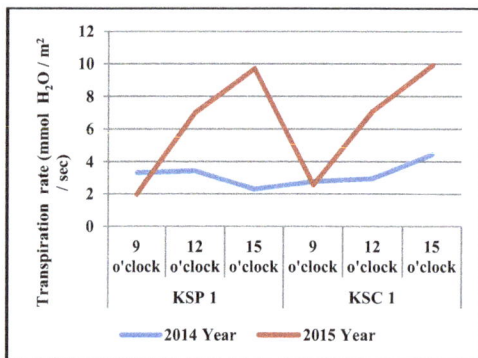

Figure 6. Diurnal variation of transpiration sweet potato plant, depending on the variety and the crop year

In 2014 when the amount of degrees the temperature was lower than the annual average and the amount of rainfall exceeded the annual average, the variety Pumpkin (KSP1) recorded higher values of diurnal

variation of leaf transpiration tests carried out at 9:12 and at 15, variety Chestnut (KSC1) achieved the highest value of diurnal variation of leaf tanspiration. In 2015 when the amount of degrees the temperature was higher than the annual average and the amount of rainfall was lower than the annual average, the variety Chestnut (KSC1) recorded higher values of diurnal variation of leaf transpiration tests carried out in all 3 points of the day. This indicates that the variety Chestnut (KSC1) has the capacity to behave better in drought conditions than the variety Pumpkin (KSP1). Climatic factors have intensified leaf transpiration which recorded high values ranging from 1.97 to 9.7 mmol H_2O / m^2 / s variety Pumpkin (KSP1) and values between 2.58 to 9.93 mmol H_2O / m^2 / s variety Chestnut (KSC1). Both varieties to leaf transpiration maximum intensity registered between 12 to 15 hours and action when stress factors (drought atmospheric and pedological drought) was maximum.

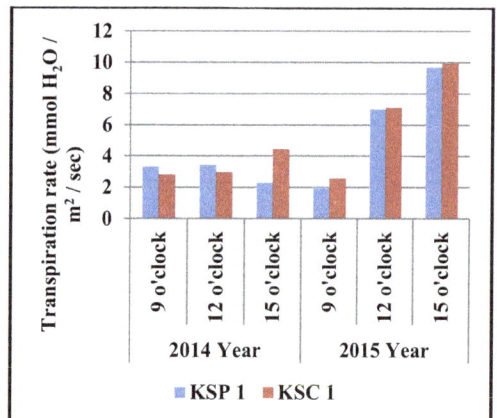

Figure 7. Diurnal variation of transpiration sweet potato plant, depending on the crop year and variety

The climatic conditions of the three years of study have influenced both the production of sweet potato and their nutritional quality (Table 2). The best production results were obtained in terms of 2015, the variety Chestnut (KSC 1) 35467kg / ha). The lowest production was obtained in the climatic conditions of 2014 (high rainfall and temperature below the annual average) (17428kg / ha variety Pumpkin (KSP1) and 18857kg / ha variety Chestnut (KSC 1).

Nutritional quality presented values differentiated the two varieties studied climate conditions of the three years.

The content of total solids of higher values under the years 2014 and 2015 (38.50% variety Pumpkin and 39.93% for the variety Chestnut), when rainfall was higher than 2013.

If the soil temperature is high dry substance can be lost through excessive breathing. In a warm season but soil moisture, dry substance will remain high due to the reduction in the intensity of process breathing.

The amount of soluble carbohydrates and vitamin C to higher values also in terms of 2014 (10.12% carbohydrate variety Pumpkin and 16.72 mg / 100g fresh substance, vitamin C variety Chestnut).

The starch content presented similar values in the years 2014 and 2015.

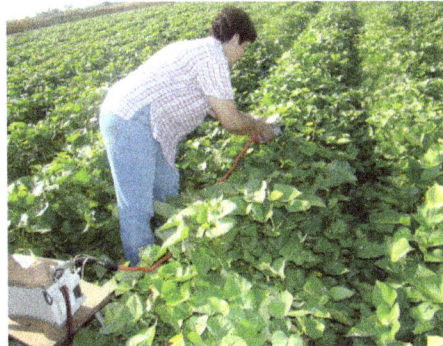

Figure 8. Determinations of plant physiology sweet potato

Table 2. Results on production of sweet potato tubers and their biochemical composition, depending on the variety and the crop year

Variety	Year	Tuber production (Kg/ha)	Total dry matter (%)	Water (%)	Soluble carbohydrates (%)	Starch (%)	C vitamin (mg / 100g fresh substance)
Pumpkin (KSP 1)	2013	23864	37.44	62.56	8.38	-	10.56
	2014	17428	38.50	61.50	10.12	12.70	11.44
	2015	33300	38.41	61.59	8.33	12.55	7.99
Chestnut (KSC 1)	2013	30176	35.44	64.56	7.97	-	13.20
	2014	18857	36.55	66.95	10.04	12.60	16.72
	2015	35467	38.93	61.07	8.09	12.69	10.56

The results obtained in the production of sweet potato tubers during 2013-2015, highlights a positive correlation with the amount of degrees of temperature recorded during the growing season (May-September) (Figure 9). This function connexion is more closely to KSC1, which is delayed compared to the variety Pumpkin 5-7 days, production is greatly influenced by climatic conditions during baking.

Production recorded in both varieties was negatively correlated with the amount of rainfall recorded during the growing of the plant (Figure 10). The climatic conditions in 2014, particularly abundant rainfall, were unfavorable to plant productivity, due to the growth of lush foliage at the expense of root tubers process;

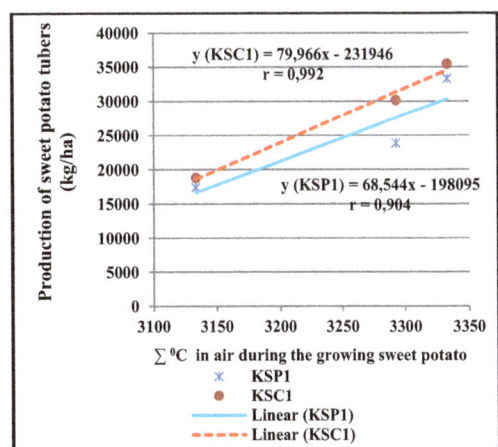

Figure 9. Relationship between production of tubers and the amount of degrees of temperature recorded during the growing season of sweet potato

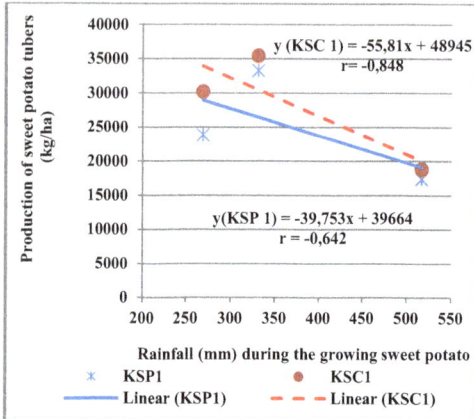

Figure 10. The relationship between production of tubers and rainfall recorded during the growing sweet potato

Figure 11. Sweet potato in the vegetation phase of and phase of tubers

CONCLUSIONS

Research conducted during 2013-2015 at the Research - Development Center for Agricultural Plants on Sands, Dabuleni, highlights a favorable microclimate for plant growth and development of sweet potato (*Ipomoea batatas*).

The values recorded physiological indices sweet potato grown in climatic conditions in the sandy soils of southern Oltenia showed that it easily adapts to the conditions of excess heat here is a plant heat-loving and light. Noting that the rate of photosynthesis in this species remains high throughout the day compared to other species (watermelons, peppers, eggplant), where the rate of photosynthesis decreases sharply from 12 to 15 hours under direct stress factors.

Production made sweet potato varieties Pumpkin (KSP 1) and Chestnut (KSC 1), studied under sandy soils ranged 17428kg / ha and 35467 kg / ha depending on the crop and cultivated variety.

Production of sweet potato tubers obtained from the period 2013-2015, is positively correlated with the amount of degrees of temperature and negatively with rainfall recorded during the growing season (May-September).

Nutritional quality presented values differentiated to the two varieties studied, according to the year of culture.

Given the level of production that was achieved in sweet potato grown in the period 2013 - 2015 is necessary to continue research on introduction in culture in the sandy soils of Romania sweet potato (*Ipomoea batatas*).

REFERENCES

Burri B.J., 2011. Evaluating Sweet Potato as an Intervention Food to Prevent Vitamin A Deficiency. Comprehensive Reviews in Food Science and Food Safety, Volume 10, Issue 2, pages 118–130.

Ciofu R., 2005. 2005, Sweet potatoes, Horticulture, nr.5-6, p 6-8.

Cioloca M., Nistor A., Chiru N., Popa M., 2013. Sweet potato - an alternative under climate change. Results of multiplication in vitro. Potato in Romania, publication of technical information for potato growers, Vol 22. No. 1: 50-53.

Duke A. J., 1983. Handbook of Energy Crops. Unpublished (Sweet potato, From Wikipedia, the free encyclopedia)

Kareem I., 2013. Fertilizer Treatment Effects on Yield and Quality Parameters of Sweet potato (Ipomoea batatas), 2013. Research Journal of Chemical and Environmental Sciences Res. J. Chem. Env. Sci., Volume 1 Issue 3 (August 2013): 40-49.

Ladokum O.A., Aderemi F.A., Tewe O.O., 2007. Sweet potato as a feed resource for layer production in Nigeria. African Crop Science onference Proceeding 8: 585-588

Muşat C., 2010, Research on some technological defining sequence for culture of sweet potatoes (Ipomoea batatas Poir.), Annals of Univ. Craiova, Series: Biology, Horticulture, Agricultural Products Processing Technology, Environmental Engineering. Vol. XV (XLXI) - 2010, 354-359.

Wariboko C., Ogidi I. A., 2014. Evaluation of the performance of improved sweet potato (*Ipomoea batatas* L. LAM) varieties in Bayelsa State, Nigeria. African Journal of Environmental Science and Technology. Vol. 8(1), 48-53.

4

THE INCIDENCE AND PREVALENCE OF ROOT-KNOT NEMATODE SPECIES (*MELOIDOGYNE* SPP.) ASSOCIATED WITH DIFFERENT DICOTYLEDONS ORIGINATED FROM TWO VEGETABLE CROPPED AREAS, VĂRĂŞTI (GIURGIU), AND BĂLENI (DÂMBOVIŢA)

Leonard BOROŞ[1]**, Tatiana Eugenia ŞESAN**[2]**, Mariana Carmen CHIFIRIUC**[2]**,
Ionela DOBRIN**[3]**, Beatrice IACOMI**[3]**, Claudia COSTACHE**[4]

[1]Phytosanitary Unit, Regional Laboratory of Nematology, 47 Lânii Street, 500465, Braş ov, Romania, Email: miksogenis@yahoo.gr

[2] University Bucharest, Faculty of Biology, [2a]Research Institute of the University of Bucharest – ICUB, Spl. Independenţei 91-95, Bucharest, Romania
E-mail: tatianasesan@yahoo.com, carmen_balotescu@yahoo.com

[3] University of Agronomic Sciences and Veterinary Medicine of Bucharest, 59 Blvd. Mărăşti, Bucharest, 011464, Romania, E-mail: ioneladobrin@gmail.com, b.iacomi@yahoo.fr

[4] Central Phytosanitary Laboratory, 11 Voluntari Blv.,077190 Voluntari, Ilfov, Romania, E-mail: claudia.costache@lccf.ro

Corresponding author email: miksogenis@yahoo.gr

Abstract

Although the vegetable fields is reduced in Romania, the production losses remain high and this concern is not owed only to fungui, viruses, bacteria or insects, but also to nematodes, these last organisms being less known and acknowledged. Meloidogyne incognita (Kofoid & White, 1919) and Meloidogyne hapla (Chitwood, 1949) are considered important parasitic nematodes, but quite little studied in Romania, for different species of dicotyledon vegetables and this article demonstrates and compares the development and reproduction of these two species related to one of the most important vegetable cultures from the economic point of view. It seems that the Meloidogyne hapla species especially prefer the species from the Apiaceae family, unlike those belonging to Meloidogyne incognita which develop a more intense shock on Brassicaceae and Solonaceae botanic family species. However, the eggs masses detected for both root-knot nematodes species show a unitary type of distribution at the surface of the of roots cortical area of all analyzed vegetable species. A contamination with the Meloidogyne incognita species, unusually high, has been noticed for the first time in our country in Brassica oleracea species.The diagnosis through biomorphometry on semi-permanent microscopic preparations, completed with the molecular biology techniques (restriction fragment length polymorphism - $RFLP_s$) led to the conclusion that the twocategories of diagnosis methods can be considered as being complementary.

Key words: diagnosis, galles, larves, root-knot nematodes (Meloidogyne spp.).

INTRODUCTION

The genus *Meloidogyne* Göldi 1892 or the root-knot nematodes (RKN) consists of sedentary, polyphagous root endoparasites (Sharon *et al.* 2007). More than 100 species have been reported worldwide (Karssen & Moens 2013). Although nematodes from the *Meloidogyne* family were reported for the first time in Romania, at the end of the 40s, under the name of "root worms" or *Heterodera maroni* Cornu (Manolache *et al.* 1949), the studies elaborated along the time, focused on the control issues,

attack methods and to the produced damages, and less on the attacked host plants. At the same time, there are little references regarding the share of *Meloidogyne* species, in the large fields cultivated with vegetables in our country. The nematode is very harmful causing essential losses especially in the tropical and subtropical areas. Unfortunately, our country is not avoided by their more or less aggressive attack, despite its location in the temperate area.

The root-knot nematodes (RKN) are making the object of large studies regarding the aggression level of the attacks and the management of such attacks, having as final purpose, to reduce the application of chemical or non-chemical (in tests) nematocides. In order to reach this, a new approach is needed, based on a correct identification of *Meloidogyne* species.

Therefore, the objectives of this study were to establish the incidence and the prevalence of RKN species, in two important vegetable fields, the vegetables host associated with the attack of these nematodes, the density of the last ones, as well as the combination of the classic diagnosis methods with those of molecular biology.

MATERIALS AND METHODS

The nature of the samples was constituted of soil and roots resulted from two important vegetables fields, Vărăşti, respectively Băleni. Vărăşti commune is located in the passage of Săbarului valley, at the east limit of Giurgiu County (South-Eastern Romania), at 30 km from Bucharest. Băleni commune is located in the south side of Dâmboviţa County (Central-Southern Romania) situated at a distance of around 20 km from Târgovişte Municipality. Both areas are under the incidence of the temperate-continental climate, characterized by very hot summers, moderate rainfalls and not so cold winters, with rare winter blizzards and frequent warming periods.

The biotic and abiotic diversity allows the cultivation of a large variety of vegetables, on protected fields and especially on agrarian fields, providing an important amount from the vegetables production of the country.

However, none of both fields was avoided by the pest attack, including the root-knot nematodes (RKN).

The samples (mixture of soil and vegetables roots) were collected within April 2014 – November 2014 and the identification of root-knot nematodes species belonging to *Meloidogyne* family was made on stages.

There were mainly chosen plants with low fructification, dwarfing phenomena and different levels of wilting.

The soil was collected from a depth of 15-20 cm, using a hand shovel, following a zig-zag model. The adherent soil on roots collected from 11 micro-farms (6 from Băleni - 5ha, respectively 5 micro-farms from Vărăşti – 6ha) was carefully shacked, attentively observing the roots (presence/absence of galls). Then two samples of 1 kg soil and roots/ha were collected. The samples were divided in sub-samples of around 200 g. This sub-division was necessary for a better laboratory processing.

The samples were stored in polyethylene bags at temperatures of 7 - 10°C, until their processing.

The storage of the last ones was performed on stages for a period of 5-7 days.

The nematodes extraction from soil was performed using the Cobb's method, through sieving and decantation and afterwards using the Baermann modified method (Southey, 1985). The nematodes were collected in aqueous suspension during no more than three days, numbered on counting dish, using a binocular stereomicroscope (Leica MZ95) and the density was established as number of nematodes on 200 g soil.

To establish the density in roots, they were shacked to remove the adherent soil, carefully washed and cut in pieces of 1-2 cm.

The nematodes extraction was performed placing the roots (aprox 10 g) in hatching chamber to produce juveniles hatching from the eggs (McKenry and Roberts, 1985). After five days, the nematodes were numbered under a binocular stereomicroscope.

The perineal pattern was placed in glycerine, after the females were dissected from the roots galls to be identified (Jepson, 1987).The nematodes were killed in water at 70°C, fixed in TAF and placed in glycerine, on a glass slide sealed with paraffin ring, for identification, using a stereomicroscope Leica DMLB with camera Leica DC300 and Leica DFC295 image-processing software. The severity of the attack on roots was evaluated on a scale from 0 to 5, as it follows: 1= 1-2 galls; 2= 3-10 galls; 3= 11-30 galls; 4= 31-100 galls; 5= over 100 galls (Taylor and Sasser, 1978).

The reception of the samples, their storage, the extraction of the nematodes, their preliminary examination, the microscopic preparations and the bio-morphometric identification were performed at the Regional Laboratory of Nematology-Braşov while the bio-morphometric observations of species confirmation and diagnosis through molecular biology were performed at the Central Phytosanitary Laboratory–Bucureşti.

The RKN incidence for each culture (host plant), for each vegetable field, as well as the global (total) incidence, were calculated using the formula below (Hussain, 2012):

$$\text{Incidence } (\%) = \frac{\text{Total number of infected plants}}{\text{Total number of observed plants}} \times 100$$

The prevalence for each vegetable field was calculated using the formula below (Hussain, 2012):

$$\text{Prevalence } (\%) = \frac{\text{Total number of fields with root-knot nemaodes}}{\text{Total number of fields surveyed}} \times 100$$

The incidence of each RKN species (occurrence) for each vegetable field was calculated using the formula (Norton, 1978):

$$\text{Occurrence of specie } (\%) = \frac{\text{Number of sample within species}}{\text{Total number of samples observed}} \times 100$$

In terms of molecular biology analysis there were used juveniles of the two species, following the steps below. Therefore, DNA extraction was carried out using ten juveniles collected directly from samples.

The juveniles were crushed between a glass slide and the cover slip by gentle pressure.

The extract was recovered with 20 μL of lysis buffer (10 mM Tris pH = 8.8, 1 mM EDTA, 1% Triton X-100, 100 mg/mL proteinase K) incubated at 60°C for 1 h, then at 95°C for 10 min (Ibrahim et al., 1994). The ITS regions of rDNA were amplified using the forward primer 18S 5'-TTG ATT ACG TCC CTG CCC TTT-3' and reverse primer 26S 5'-TTT CAC TCG CCG TTA CTA AGG-3' (Vrain et al., 1992).The PCR mixture (total volume 25 μL) contained 1x buffer enzyme, 1, 5 mM $MgCl_2$, 0.6 μM of each primer, 1 U Taq DNA polymerase (MP Biomedicals), 0.2mM dNTPs (MP Biomedicals) and 5 μL DNA extract

AMasterCycler Pro S (Eppendorf) was used for amplification, and the reaction consisted of a denaturation step at 94°C for 15 min followed by 45 cycles at 94°C for 15 s, 58°C for 30 s,

72°C for 1 min, and a final extension step of 10 min 72°C. Following PCR, 10 μL of the amplified product was analysed by electrophoresis in a 1% agarose gel.Amplified DNA was digested with DraI and RsaI restriction endonucleases (Fermentas and Promega) using an aliquot of 5 μL of the PCR product and 5 U of each enzyme, according to the manufacturer's instructions. Species-specific ITS–RFLP profiles for Meloidogyne were generated using these two restriction enzymes (Zijlstra et al., 1995).

Fragments were resolved by electrophoresis in 1,5-2% agarose gel. Data analysis was performed using GENi (Syngene) and 100 bp DNA Ladder (GeneRuler, Fermentas) as a molecular size marker.

Analyses of nematodes density were carried out using SPSS Statistics (Statistical Package for the Social Sciences, available online at https://statistics.laerd.com/spss-tutorials/one-way-anova-using-spss-statistics.php) and the Standard error calculation (available online at http://www.investopedia.com/terms/s/standard-error.asp).

RESULTS AND DISCUSSIONS

The most important vegetable cultures from Băleni (Dâmbovița County), namely Vărăşti (Giurgiu County) were studied: beet, celery, parsnip, carrot, parsley, lettuce, broccoli, cauliflower, cucumber, vegetable-marrow, tomato and pepper.

Among the 120 analysed samples, the majority composed of a mixture of soil and roots, 105 samples were positive for RKN.

From the total number of samples which were examined, from both regions, it resulted that the global incidence is 87.5 % infested samples with RKN, but there are differences concerning the incidence reported to different vegetables species.

Therefore, the incidence from Băleni area starts from 50% for *Lactuca sativa* and *Cucurbita pepo*, achieving the maximum percentage in case of *Apium graveolens* (Figure 1), *Pastinaca sativa* (Figure 2), *Brassica oleracea* and *Brassica oleracea* var. *botrytis* (Table 1).

From Vărăşti area the incidence starts from 50% for *Beta vulgaris* achieving the maxim percentage to *Daucus carota* and *Petroselinum crispum*, among others (Table 2).

In Vărăşti area, the incidence for all positive (+) samples of vegetables is 88.33 %, unlike Băleni area, where this incidence was 86.66%. Although these differences between the two vegetable areas are not too significant, it is to be noticed the RKN incidence of 100% to the family members of *Brassicaceae* from Băleni area, along with a significant deformation of the roots.

Figure 2. Infested roots – *Pastinaca sativa*

This deformation is due the galls. Each gall usually contains three to six giant cells, which are due to substances contained in the "saliva" secreted by the nematode in the giant cells during feeding. The giant cell crush xylem elements already present but degenerate when nematodes cease to feed or die. In the early stages of gall development the cortical cell enlarge in size, later, they also divide rapidly. Frequently other parasites can easily attack the weakened root tissues and the hypertrophied, undifferentiated cells of the galls (Agrios, 2005).

Figure 1. Infested roots – *Apium graveolens*

Table 1. The incidence of RKN (%) for crop host at Băleni – Dâmboviţa
(+ present, - absent, RKN root – knot nematode)

Botanic family	Crop host	Total soil samples	+	-	Incidence RKN %
Amaranthaceae	Beta vulgaris	4	3	1	75
Apiaceae	Apium graveolens	4	4	-	100
	Pastinaca sativa	6	6	-	100
	Daucus carota	7	6	1	85.7
	Petroselinum crispum	7	6	1	85.7
Asteraceae	Lactuca sativa	2	1	1	50
Brassicaceae	Brassica oleracea	5	5	-	100
	Brassica oleracea var. botrytis	7	7	-	100
Cucurbitaceae	Cucumis sativus	3	2	1	66.66
	Cucurbita pepo	2	1	1	50
Solanaceae	Lycopersicon esculentum	11	10	1	90.90
	Capsicum annuum	2	1	1	50
Incidence of RKN/area 86.66 %					

Table 2. The incidence of RKN (%) for crop host at Vărăşti – Giurgiu
(+ present, - absent, RKN root – knot nematode)

Botanic family	Crop host	Total soil samples	+	-	Incidence RKN %
Amaranthaceae	Beta vulgaris	2	1	1	50
Apiaceae	Apium graveolens	3	3	-	100
	Pastinaca sativa	8	7	1	87.5
	Daucus carota	7	7	-	100
	Petroselinum crispum	5	5	-	100
Asteraceae	Lactuca sativa	1	1	-	100
Brassicaceae	Brassica oleracea	3	2	1	66.66
	Brassica oleracea var. botrytis	3	2	1	66.66
Cucurbitaceae	Cucumis sativus	4	3	1	75
	Cucurbita pepo	2	1	1	50
Solanaceae	Lycopersicon esculentum	13	13	-	100
	Capsicum annuum	9	8	1	88.88
Incidence of RKN/area 88.33 %					

The incidence of each species of nematodes (occurrence) in different cultures of host-plants, underlines the fact that Meloidogyne incognita species is predilected to Solanaceae, Cucurbitaceae and to Brassicaceae (Table 3). The Meloidogyne hapla species grow better on species of root vegetables, in other words species of vegetables of the Apiaceae family (Table 4). Occurrence of species Meloidogyne hapla at Băleni area is predominant (50%) unlike the species Meloidogyne incognita (36.66%). Occurrence of species Meloidogyne incognita is higher at Vărăş ti area (53.33%) while Meloidogyne hapla was found in less

than 50% (31.66%). The index of the galls is a sign of the nematodes presence or absence in roots and implicitly in soil. This index indicates the severity of the attack. Therefore, this severity is maximum (5) to Brassicaceae family (Figure 3) from Băleni while the index of the galls is raised (4) to Apiaceae family (Figure 4) and Solanaceae families from Vărăşti.
The mixed RKN population are relatively common in the vegetables plantations, only that in our study they were found only to Lycopersycom esculentum specie from Băleni area.

Table 3. The incidence of each RKN species (occurrence) and the index of the galls at
Băleni - Dâmboviţa area (*M.i. = Meloidogyne incognita, M.h. = Meloidogyne hapla*)

Botanic family	Crop host	M.i.	M.h.	Non detected	Gall index
Amaranthaceae	*Beta vulgaris*	-	3	1	2
Apiaceae	*Apium graveolens*	-	4	-	4
	Pastinaca sativa	-	6	-	2
	Daucus carota	-	6	1	3
	Petroselinum crispum	-	6	1	4
Asteraceae	*Lactuca sativa*	-	1	1	1
Brassicaceae	*Brassica oleracea*	5	-	-	5
	Brassica oleracea var. *botrytis*	7	-	-	5
Cucurbitaceae	*Cucumis sativus*	2	-	1	3
	Cucurbita pepo	-	1	1	3
Solanaceae	*Lycopersicon esculentum*	7	3	1	4
	Capsicum annuum	1	-	1	2
	Occurrence RKN (%)	36.66	50	13.34	

Table 4. The incidence of each RKN species (occurrence) and the index of the galls at
Vărăşti - Giurgiu area (*M.i. = Meloidogyne incognita, M.h. = Meloidogyne hapla*)

Botanic family	Crop host	M.i.	M.h.	Non detected	Gall index
Amaranthaceae	*Beta vulgaris*	1	-	1	1
Apiaceae	*Apium graveolens*	3	-	-	3
	Pastinaca sativa	-	7	1	4
	Daucus carota	-	7	-	4
	Petroselinum crispum	-	5	-	4
Asteraceae	*Lactuca sativa*	1	-	-	2
Brassicaceae	*Brassica oleracea*	2	-	1	1
	Brassica oleracea var. *botrytis*	2	-	1	1
Cucurbitaceae	*Cucumis sativus*	3	-	1	3
	Cucurbita pepo	1	-	1	3
Solanaceae	*Lycopersicon esculentum*	13	-	-	4
	Capsicum annuum	8	-	1	4
	Occurrence RKN (%)	53.33	31.66	15.01	

Figure 3. Infested roots – *Brassicaceae (Brassica oleracea)*

Figure 4. Infested roots – *Apiaceae (Daucus carota)*

The prevalence was of 100% in both vegetable fields, so that all studied fields were infested with root – knot nematodes.

At least 40 perineal patterns from each vegetable field were examined for better identification accuracy (Figure 5 and 6).

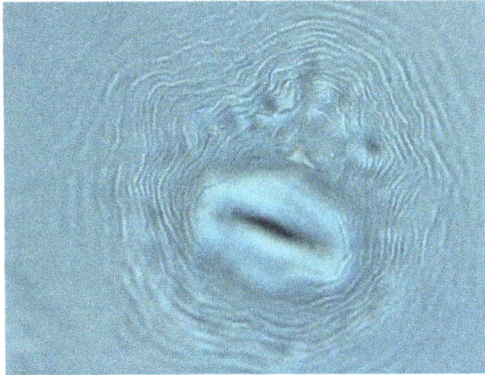

Figure 5. *Meloidogyne incognita* – female perineal pattern (100x magnification). Scale bar =20µm

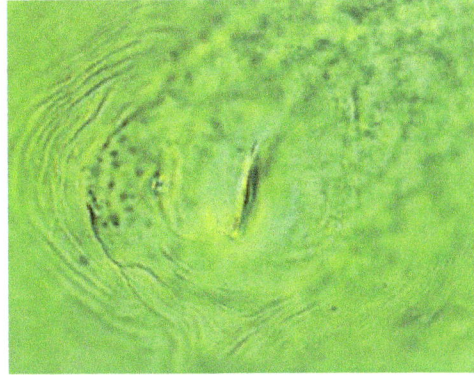

Figure 6. *Meloidogyne hapla* - female perineal pattern (100x magnification). Scale bar =20µm

Statistical analysis used fuctions Mean and ANOVA belonging program SPSS. Mean number of nematodes on roots was significantly higher at Băleni (11/10 g roots) compared to 8.25/10 g roots al Vărăş ti (Sig. 0.0006). At Băleni the total number of nematodes (soil+roots) was significantly higher compared to Vărăş ti (16.89)(Table 5).

At Băleni area the highest total mean number of nematodes was found at *Brassica oleracea* var. *botrytis* (134.8).
The highest total mean number of nematodes at Vărăş ti area was at *Cucurbita pepo* of which 42.2 nematodes/200 g soil and 51.4 nematodes/10 g roots (Table 6).

Table 5. Nematode population density per 200 g soil + 10 g roots of diseased crop host at Băleni (± Standard error)

Crop host	Positive samples (+)	Mean number of nematodes ±Standard error		
		200 g soil	10 g roots	Total
Beta vulgaris	3	6.53±0.40	6.73±0.34	13.27±0.61
Apium graveolens	4	9.40±0.76	10.75±0.48	20.15±0.97
Pastinaca sativa	6	3.37±0.34	3.27±0.35	6.63±0.49
Daucus carota	6	6.63±0.34	5.07±0.41	11.70±0.55
Petroselinum crispum	6	6.80±0.34	6.67±0.38	13.47±0.55
Lactuca sativa	1	2.40±0.51	10.60±1.81	13.00±1.38
Brassica oleracea	5	32.08±0.75	35.60±0.92	67.68±0.95
Brassica oleracea var. *botrytis*	7	61.60±1.03	73.20±1.46	134.80±1.07
Cucumis sativus	2	21.40±1.71	30.80±1	52.20±2.16
Cucurbita pepo	1	21±2.05	26.00±1.52	47.00±1.70
Lycopersicon esculentum	10	4.38±0.28	5.12±0.40	9.50±0.47
Capsicum annuum	1	2.43±0.27	2.83±0.31	5.26±0.41
Mean at Băleni	52	9.75±0.72	11.0±0.9	20.8±1.55

Table 6. Nematode population density per 200 g soil + 10 g roots of diseased crop host at Vărăş ti (± Standard error)

Crop host	Positive sample (+)	Mean number of nematodes ±Standard error		
		200 g soil	10 g roots	Total
Beta vulgaris	1	7.00±0.63	2.20±0.49	9.20±1.02
Apium graveolens	3	8.67±0.40	12.67±0.61	21.33±0.79
Pastinaca sativa	7	8.91±0.37	8.23±0.36	17.14±0.58
Daucus carota	7	6.20±0.31	7.26±0.25	13.46±0.36
Petroselinum crispum	5	14.72±1.01	12.48±0.26	27.20±1.27
Lactuca sativa	1	21.40±0.51	19.20±1.24	40.60±1.50
Brassica oleracea	2	1.10±0.28	2.80±0.51	3.90±0.64
Brassica oleracea var. *botrytis*	2	1.03±0.19	0.37±0.12	1.40±0.22
Cucumis sativus	3	11.70±0.78	14.00±2.07	25.07±2.28
Cucurbita pepo	1	42.20±0.80	51.40±4.76	93.60±4.48
Lycopersicon esculentum	13	5.91±0.23	4.85±0.53	10.75±0.61
Capsicum annuum	8	30.60±1.60	21.10±1.53	51.70±1.86
Mean at Vărăş ti	53	8.63±0.50	8.25±0.53	16.89±0.99

Restriction fragment length polymorphisms (RFLP$_s$) method has the advantage that it distinguishes species after extraction and purification of genomic DNA, restriction digestion and visualisation of banding patterns in gel electrophoresis. The 760 bp PCR product we obtained for the amplified ITS region with 18S and 26S primers. After digestion PCR products with the two restriction enzymes , *Meloidogyne hapla* isolate showed the following restriction patterns: 380 bp with DraI and 620, 140 bp with RsaI and for *Meloidogyne incognita* isolates: 220, 200, 180, 160 bp with DraI and 760 bp with RsaI (Figure 7).

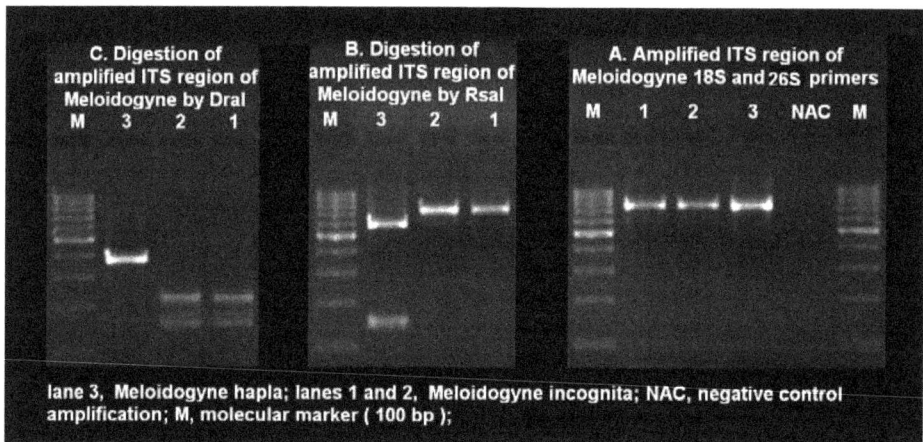

Figure 7. **A**, typical amplification by polymerase chain reaction (PCR) of 760 bp product from template of total DNA extracted from juveniles of *Meloidogyne hapla* and *Meloidogyne incognita* ; **B, C,** size of DNA fragments (bp) obtained after restriction enzyme digestion of the 760 bp internal transcribed spacer regions of *Meloidogyne hapla* and *Meloidogyne incognita* with RsaI and DraI.

CONCLUSIONS

The present study was elaborated to identify and quantify one of the most encountered species of polyphagous phyto-parasite nematodes in our country.
To better understand the economic impact of the attacks in case, a correct identification of RKN was necessary.

Both vegetables fields are favourable to the RKN development due to the soil, the temperature and the host-plants presence, as proved by the high global incidence which reaches 87.5%.
Until present, there was no aggressive attack to *Brassica oleracea* reported in the specialty literature from our country.

In this case, the significant density of nematodes in soil and roots, probably the susceptibility of the variety, leaded to an incidence of 100% to *Brassicaceae* family in Băleni.

The preponderance of the *Meloidogyne hapla* (50%) species was observed in Băleni unlike Vărăşti where the *Meloidogyne incognita* (53.33%) species predominates.

It is to remark that as far as *Lycopersycom esculentum* (culture of major economic importance) is concerned, the occurrence of the RKN was 100% which could be due, among other factors, to the use of Monkeymaker variety, sensible to RKN attack.

An index of the galls over 100/root was reported to the members of *Brassicaeae* family (Băleni).

The density of nematodes (in soil and embedded in the roots) is relatively close to each host plants.

The difference in terms of density among different plant species are due to climate, soil type, plant varieties, which favours the survival and multiplication of RKN.

Although molecular techniques are not readily available to every diagnostician, they complement and confirm the bio-morphometrical identification, thus increasing the reliability of diagnosis.

The combination of the conventional and molecular methods represents, from our point of view, a challenge for the discovery of other species of *Meloidogyne* parasitic in the vegetable areas from different areas of the country.

REFERENCES

Agrios G.N., 2005. Plant Pathology, 5th edn. Academic Press, USA.

Holbrook C.C., Knauft D.A., Dickson D.W., 1983. A Technique for Screening Peanut for Resistance to *Meloidogyne arenaria*. Plant Disease. 67 (9): 957-958.

Hussain M. A., Mukhtar T., Kayani M. Z., Aslam M. N.,. Haque M. I., 2012. A survey of okra (*Abelmoschus esculentus*) in the Punjab province of Pakistan for the determination of prevalence, incidence and severity of root-knot disease caused by *Meloidogyne spp*. Pakistan Journal of Botany, 44 (6): 2071-207.

Ibrahim S.K., Perry R.N., Burrows P.R., Hooper D.J., 1994. Differentiation of species and populations of *Ditylenchus angustus* using a Fragment of Ribosomal DNA. Journal of Nematology, 26:412–421.

Jepson S. B., 1987. Identification of root-knot nematodes (*Meloidogyne* species). Wallingford, C.A.B. International, London.

Karssen G., Moens M., 2013. Root-knot nematodes. In: Perry RN, Moens, M. Plant Nematology, 2nd edition. CAB International, Wallingford, UK, 59-90.

Manolache C., Pain S., Săvescu A., Bucş an I., Manolache F., Hrisafi C., 1949. Situaţ ia dăunătorilor animali ai plantelor cultivate în anul 1947-1948. Seria Nouă (1): 11.

McKenry M. V., Roberts P.A., 1985. Phytonematology study guide. Univ. of California, Div. of Agri. and Natural Res. Pub.

Norton D.C., 1978. Ecology of plant parasitic nematodes.Wiley and Sons, New York.

Sharon E., Chet I., Viterbo A., Bar-Eyal M., Nagan H., Samuels G.J., Spiegel Y., 2007. Parasitism of *Trichoderma* on *Meloidogyne javanica* and role of the gelatinous matrix. European Journal of Plant Pathology,(118): 247-258.

Southey J.F., 1985. Laboratory methods for work with plant and soil nematodes. Her Majesty's stationary office, London.

Taylor A.L., Sasser J.N., 1978. Biology, identification and control of root-knot nematodes (*Meloidogyne* species). Raleigh NC, USA.

Taylor D.P., Netscher C., 1974. Improved technique for preparing perineal patterns of *Meloidogyne* spp. Nematologica, 20(2): 268-269.

Viaene N., Moens, M., 2011. Root-knot nematodes in Europe. Nematology (13): 3-16.

Vrain T.C., Wakarchuk D.A., Levesque A.C., Hamilton R.I., 1992. Intraspecific rDNA restriction fragment length polymorphism in the *Xiphinema americanum* group. Fundamental and Applied Nematology, 15: 563–573.

Zijlstra C., Lever A.E.M., Uenk B.J., Van Silf hout C.H., 1995. Differences between ITS regions of isolates of root-knot nematodes *Meloidogyne hapla* and *M. chitwoodi*. Phytopathology, 85: 1231-1237.

PM 7/41 (2) EPPO Standard . 2009, *Meloidogyne chitwoodi* and *Meloidogyne fallax*. Bulletin OEPP/EPPO 39, 5-17.

5

PRE-BREEDING FOR DIVERSIFICATION OF PRIMARY GENE POOL IN ORDER TO ENHANCE THE GENETIC PEPPER RESOURCES

Petre Marian BREZEANU[1], Creola BREZEANU[1,2], Silvica AMBARUS[1], Teodor ROBU[2], Tina Oana CRISTEA[1], Maria CALIN[1]

[1]Vegetable Research and Development Station Bacău, Calea Bîrladului No 220, Bacău, Romania
brezeanumarian@yahoo.com, creola.brezeanu@yahoo.com, silvia_ambarus@yahoo.com
[2]Ion Ionescu de la Brad" University of Agricultural Sciences and Veterinary Medicine Iasi, Faculty of Agriculture, 3, Mihail Sadoveanu Alley, Iasi, Romania
Corresponding author email: creola.brezeanu@yahoo.com

Abstract

Exploitation of new and diverse sources of variation is needed for the genetic enhancement of Capsicum annuum L. species. Variety is an important factor of production and has to be in accordance with consumer needs. This study represents a screening of physiology and biochemistry of pepper, particular the Romanian cultivars. The investigated parameters were selected because of their importance in pepper quality, with the final purpose to identify the most valuable resources to be used in breeding as staring material. In this purpose, the study focused on the phenological observation, biometrical measurements and also physiological processes that occur in fruits during their growth and development, which included the following characteristics: total dry content and water content; content in soluble glucides and titratable acidity, content in β carotene and anthocians, content in ascorbic acid, all parameters investigated are related with fruit quality. The most valuable cultivars, regarding nutritional quality were: Creola, characterised by 7.90% content in dry substance, TA 0.343 mg g^{-1} malic acid, AA 200.4 mg g^{-1}, and carotenes 23.452 mg g^{-1}; Lider, 7.80 % content in dry substance, TA 0.344 mg g^{-1} malic acid, AA 200.2 mg g^{-1} and carotenes 23.47 mg g^{-1}; Cornel content in dry substance 7.40%, TA 0.331 mg g^{-1}, AA 199.2 mg g^{-1} and carotenes 23.0 mg g^{-1}.

Key words: biometrical measurements, phenological observation, physiological investigations.

INTRODUCTION

The pepper fruits are very tasty, healthy owing to their content of biologically active chemicals with antioxidant properties. The vegetable is an abundant source of vitamin C (Buczkowska and Najda, 2002). Physiologically ripe fruits are abundant in carotenoid pigments.

Moreover, pepper is an important source of minerals for humans (Bubicz et al., 1999). Consumers have high demands in terms of color, shape, size and taste of fruit and the producer must meet these requirements, making extra productivity, precocity and resistance to pathogens. There is the need to create better varieties with higher yield and quality to satisfy a growing demand (Pérez-Grajales et al., 2009).

The narrow genetic bases of cultivars coupled with low utilization of genetic resources are the major factors limiting production and productivity globally. Wild relatives with enhanced levels of resistance/tolerance to multiple stresses provide important sources of genetic diversity for pepper improvement. Otherwise, the local population or the developed cultivars can represent valuable sources for pepper improvement in terms of nutritional quality.

Pre-breeding provides a unique opportunity, through the introgression of desirable genes from wild germplasm or into genetic backgrounds readily used by the breeders with minimum linkage drag, to overcome this. Pre-breeding activities using promising landraces, wild relatives, and popular cultivars have been initiated, in a diverse range of programs.

Demand for pepper richer in compounds like ascorbic acid, capsaicin, beta-carotene and lycopene is increasing especially because of their demonstrated antioxidant potential.

The future production of cultivated pepper depends on improving their genetics and developing new superior cultivars with traits such as nutritional quality, disease resistance, and higher yield potential. Reported levels of phytochemical variation is due in large part to various environmental conditions (abiotic and biotic stresses) acting on plants during their growth and development (Leskovar et al., 2009). None the less, continual selection of material containing higher levels of these phytochemicals is a valuable component of a breeder's program and will undoubtedly result in creation of improved germplasm consumers can eat to benefit their well-being (Crosby et al., 2007).

Significant variation in phytochemical expression within pepper fruit tissue is dependent upon several factors. Genotypic, as well as, environmental differences have both contributed to material of variable phenotypic expression (Draghici, 2014). The ultimate goal of pepper breeders is, therefore, to use knowledge and apply it in a special manner to exploit more effectively match the best genotype with its optimum environment to achieve the most desirable output.

MATERIALS AND METHODS

The experiments were conducted at the Vegetable Research and Development Station Bacau.

The investigated genotypes were grown in open field, natural conditions.

The biological material was represented by twelve Romanian cultivars, as follows: 'Splendid', 'Madalin', 'Meteorit', 'Cornel', 'Lider', 'Creola', 'Granat', 'Timpuriu de București', 'Titan', 'Rubin', 'Superb', 'Globus', all cultivated in similar experimental condition.

All fruits harvested for investigations were selected at an appropriate maturity stage and size, and were healthy and turgid.

The phenological observations and the biometrical measurements were accomplished in the experimental parcels, and involved: colour at physiological maturity, fruit's length (cm), fruit's diameter (cm), ratio length/ diameter, number of lobs, fruit's weight (g), number of fruits /plant, pulp's width (mm).

Fruit measurements were conducted on fruits to gain insight into their potential variation.

The physiological changes monitored were: the content in total dry matter, water and minerals, soluble dry matter, titratable acidity,

β carotene, anthocians, ascorbic acid, glucides, all related with nutritional quality of pepper.

Maturated peppers have been collected in the same week of ripening and were chemically analyzed.

The determination of total dry matter substance was carried out by weighing the fresh vegetal material, drying it for 24 hours at 105°C, cooling it, and then weighing again the dry vegetal material. The obtained results were expressed in percentage. The difference till 100% represents the water content.

The content in mineral elements was determined by tissue incineration at 560 °C temperature and the results were expressed in percentage.

The soluble dry matter content was determined using a refractometer method and the results were expressed as a percentage.

Titratable carotene content was extracted in petrol ether and determined using a spectrometer at $\lambda=415$ nm. The content of β carotene was expressed in mg 100^{-1} g. The anthocianic pigments were extracted in methyl alcohol + 1 % HCl and spectrophotometrically determined at 540 nm wave length.

Ascorbic acid was extracted in oxalic acid 1 %, and determined with a Nexus spectrometer (FT-IR). The quantity of ascorbic acid was expressed in mg $100g^{-1}$.The content in soluble sugar was determined by Fehling method.

RESULTS AND DISCUSSIONS

The breeding program of pepper, founded by Cardi in 1997, has managed to get results in three directions; among those is the use of local landraces. A selection for stability has permitted creation of valuable cultivars from local landraces (Herman, 2005).

Our germplasm study has a multiple approach (1) the screening of phenological and morphological aspects in order to detect the most valuable resources according to the market request: pulp's width, number of lobs, colour, shape - ratio length/ diameter, fruit's weight; (2) investigation of internal quality: total dry matter, water and minerals, soluble dry matter, titratable acidity, β carotene, anthocians, ascorbic acid, glucides, in order to distinguish the most favorable germplasm for potential release in the future.

The main fruit's characteristics investigated are presented in table 1 - round pepper (12 genotypes). The most obvious trait of interest to breeders and growers is uniformity. Because pepper are a self-pollinating crop, this has been accomplished by inbreeding peppers, while selecting for important shape, flavor, appearance and yield traits by breeders throughout the world.

Table 1. Phenological observations and biometrical measurements – round pepper

| Variety | Fruit | | | |
	Length (cm)-	Diameter (cm)	Ratio L/D	No of lobs
Splendid	7.2 ± 0.030	8.6 ± 0.02	0.83 ± 0.03	3.2 ± 0.10
Madalin	7.8 ± 0.021	7.9± 0.01	0.98 ± 0.06	3.8 ± 0.10
Meteorit	7.1 ± 0.010	7.9± 0.02	0.89 ± 0.18	2.8 ± 0.04
Cornel	7.1 ± 0.040	8.2± 0.01	0.86 ± 0.21	2.9 ±0.05
Lider	7.0 ± 0.031	8.2± 0.02	0.85 ± 0.05	2.9 ± 0.17
Creola	7.9 ± 0.012	8.8± 0.01	0.89 ± 0.02	3.6 ± 0.12
Granat	7.1 ± 0.013	7.6± 0.02	0.93 ± 0.04	2.6 ± 0.9
Globus	5.6 ± 0.012	5.9 ± 0.02	0.94 ± 0.04	2.1 ± 0.14
Titan	7.4 ± 0.011	8.7± 0.02	0.85± 0.02	3.2 ± 0.52
Rubin	5.2 ± 0.020	5.9± 0.03	0.88± 0.01	2.8 ±0.42
Superb	5.9 ± 0.050	6.2± 0.04	0.95 ±0.04	2.7 ± 0.37
Timpuriu de Bucureşti	5.3 ± 0.040	6.0 ± 0.05	0.88 ± 0.05	2.8 ±0.12
Average	6.7	7.5	0.89	2.70
Standard deviation	2.56	1.33	1.6	0.29
LSD 0.05	0.98	0.72	0.85	0.20

Table 2. Phenological investigation at round pepper fruits (weight, pulp's width, number fruits/plant)

| Variety | Shape | Colour at physiological maturity | Fruit | | |
			Weight -g-	Pulp's width -mm-	Number of fruits /plant
Splendid	round - flattened	Red	70.8 ± 2.5	7.8 ± 0.090	9.2 ± 0.75
Madalin	round - flattened	red-carmine	80.6 ± 3.9	9.1 ± 0.047	6.8 ± 0.82
Meteorit	round	dark red	85.7 ±2.23	9.4 ± 0.180	7.3± 0.75
Cornel	round	Red	90.8 ± 2.4	10.2 ± 0.550	8.5 ± 0.80
Lider	round - flattened	red-carmine	75.4 ± 1.55	9.5 ± 0.600	10.3 ± 0.98
Creola	round	shiny red	215 ±4.20	12.9 ±0.190	12.8 ± 1.21
Granat	round	dark red	110 ±1.55	9.8 ±0.500	9.1 ± 1.23
Globus	globular	red-carmine	98 ± 2.20	11.2 ± 0.420	15.2 ± 1.55
Titan	spherical	dark red	180 ± 1.95	11.6 ± 0.370	6.2 ± 1.21
Rubin	round	red	90.5 ±2.45	8.9 ± 0.620	9.4 ± 1.6
Superb	round - flattened	red	97.4 ± 2.33	8.2 ± 0.550	-
Timpuriu de Bucureşti	globular	red	90.9± 1.9	8.8 ± 0.230	11.9 ± 0.66
Average			107.09	9.78	8.89
Standard deviation			85.8	1.02	2.66
LSD 0.05			30.26	0.45	0.98

Some of the most important features related with visual quality of pepper are the external color, weight of fruit, shape and pulp' weight. The tremendous variability regarding shape, color and weight of bell pepper fruits is totally different, being lower in case of round pepper germplasm. In any case, the fruit of round pepper must be fully red to be acceptable for processing and desirable for their decorative color and the flavor they impart to processed food. The consumers prefer a red fruit in a ready-to-eat stage with an attractive appearance, a crisp texture and have a specific flavor.

In our collection, the shape varied from round flattened to round, globular, with an average of ration length / diameter 0.89 (Table 1 and 2). Comparing the length and diameter of investigated genotypes, we observed that eight genotypes registered large fruits with length more than 7 cm and diameter more than 7.6 cm. The smallest fruits regarding length and diameter values were harvested from genotype 'Rubin", 5.2 cm length and 5.9 cm diameter.

The variation of total number of lobs was between 2.1 and 3.8 with an average of 2.7. Table 2 presents the variability of fruit weight, pulp width and number of fruit per plant. The value of fruit weight varied from 70.8 g ('Splendid'), to 180 g ('Titan') and 215 g ('Creola'). The heaviest fruits of 'Creola' genotype, registered the highest value of pulp's width (12.9 mm).

The shape, the size and the fruit's weight are important parameters of yield and quality as follows: fruit's weight cumulated with number of fruits per plant, density, etc. - especially in case of yield potential estimation and fruit's weight, pulp's width, fruit's shape, and fruit's size for establishment of crop use - fresh consumption or as raw material, in food industry.

Figures 1 to 5 presents the results of the analyses focused toward the determination of the chemical composition of 12 genotypes of round pepper. The size and the quality of mature peppers were determined by interaction of genotype x environmental climate. One of the indicators of size and round pepper quality is represented by the accumulation of water and total dry matter.

The total dry matter (TDM) content varied from 7% to 7.9 %. The highest content was registered in fruits of 'Lider" and "Creola" genotypes, 7.9%, respectively 7.4%. One of the factors affecting the production of plant biomass is the concentration of mineral elements.

Regarding mineral accumulation, the total amount of minerals in round pepper fruits varied in limits of 0.09%, from 0.46% at 'Spelendid', 'Madalin', 'Titan', 'Rubin' to 0.55% at 'Globus" genotype (Figure 1).

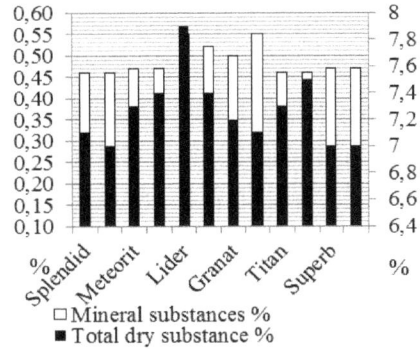

Figure 1. Total dry mater and mineral accumulation

Fruit quality and consumer acceptability in round pepper are strongly related with the pigments concentration, soluble solids content (TSS) titratable acidity (TA) and ascorbic acid (AA) in the ripened fruits. During ripening process, some substances of important nutritional quality, particularly vitamin C and carotenoids are accumulated in large quantities (Navarro et al., 2006).

Ascorbic acid, known as vitamin C, needs to be consumed via food or medicine, as it is not produced in the human organism (Manela-Azulay et al., 2003). The levels of vitamin C are variable and may be affected by maturity, genotype and processing. Ascorbic acid is the least complex vitamin found in plants and is synthesized from glucose or some other simple carbohydrate (Kays, 1991).

According to our study, the fully ripened round pepper fruits have the highest levels of ascorbic acid (AA), the titratable acidity and the total soluble solids also.

The round pepper, totally maturated represents an important source of vitamin C for human consumption, presenting values of higher than 190 mg 100 g^{-1} ('Meteorit'), till 200.4 mg 100 g^{-1} ('Creola'). Ascorbic acid (AA) concentrations were highly variable among the accessions (Figure 2).

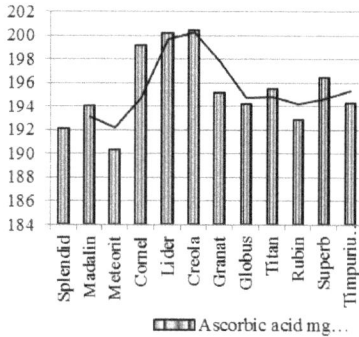

Figure 2. Ascorbic acid in maturated fruits

Our results highlight the antioxidant potential of round pepper species, stronger than those of mango (84 mg 100 g^{-1}), guava (67 mg 100 g^{-1}) and orange (40 mg 100 g^{-1}), Toda Fruta, 2004). Round pepper is one of the most important sources of vitamin C, and also the β-carotene content, the predecessor of vitamin A. The content of β-carotene was very high in most investigated varieties and fluctuates from 18.96 mg 100g^{-1} ('Meteorit') to 23.45 and 23.47 mg 100g^{-1} at 'Creola', respectively 'Lider'. This compound increased dramatically in mature colored fruit for all lines tested.

Carotenoids concentrations varied tremendously among the germplasm accessions. Six accessions registered low β – carotene content, (under 20 mg 100g^{-1}), and other six genotypes were characterized by extremely high (above 20 mg 100g^{-1}) total carotenoids at the maturated stage (Figure 3).

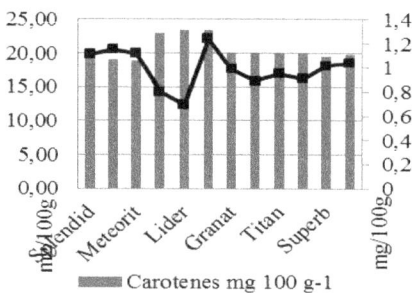

Figure 3. Pigments accumulation in pepper fruits

Investigated peppers are a good source of vitamin C and carotenoids that are important nutritional antioxidants found in the human diet. Many authors reported increase of carotenoids, during development of pepper fruits. In general, ripening of pepper fruits is strongly associated with carotenoids accumulation (Marcus et al., 1999). Anthocians synthesis is also responsible for color of maturated fruits.

The average of the anthocians content was 0.992 mg 100g^{-1}, with a variation form 0.69 mg 100g^{-1} (at 'Lider') to 1.24 mg 100g^{-1} at 'Creola' (Figure 4).

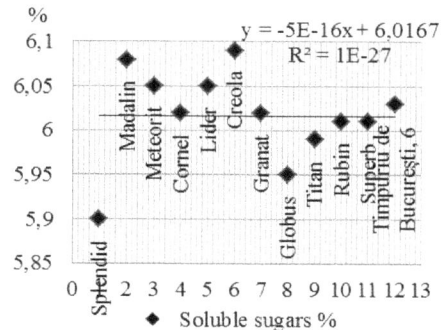

Figure 4. Soluble sugars content

The genotype 'Creola' is distinguished by its highest content in anthocians and also in β-carotene. Significant accumulation of soluble sugars during pepper fruit ripening was confirmed by findings of (Howard et al., 2000). In our study the range of soluble sugars varied from 5.90% at fruits of 'Splendid" cultivar to 6.90 % at 'Creola'.

Titratable acidity (TA) as well as TSS are commonly measured to give an overview of pepper maturity at harvest, and are used for harvest scheduling. TA indicates the total amount of organic acids. It is already known the fact that titratable acidity (TA) of the round peppers is increased with ripening, while during the ripening process the metabolic reactions increase, increasing the concentration of organic acids involved in the Krebs cycle.

Apart from this, these acids make up the energetic reserves and the metabolic reactions that involve the synthesis of pigments, enzymes and other materials and the degradation of pectins and celluloses, which are essential for the ripening process. Acidity is important for flavor balance. Our study shows a small variation in case of titratable acidity (TA) of investigated genotypes from 0.330 mg g^{-1} malic acid ('Madalin' and 'Superb') to 0.334 mg g^{-1} at ('Lider') (Fig.5). The total soluble solids

(TSS) increased as ripening of the fruit increased due to the greater degradation or biosynthesis of the polysaccharides and the accumulation of sugars.

The metabolic processes related to the advance of ripening, probably due to disassociation of some molecules and structural enzymes in soluble compounds, directly influence the levels of total soluble solids, where fruits in advanced stages of ripening present the highest levels of soluble solids (Lyon et al., 1992).

Regarding the content of total soluble solids (TSS), the variation was form 7.1 % at 'Superb', 'Globus', 'Cornel', and 'Madalin' to 8.3 % at 'Creola'.

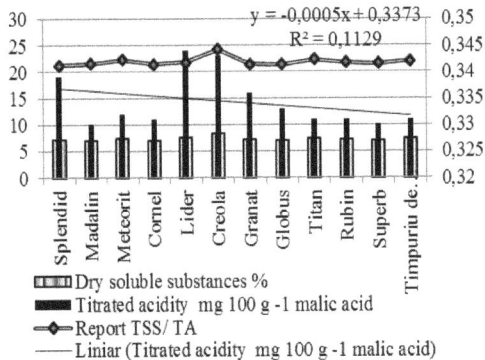

Figure 5. TTS, TA and report between

As TSS, TA and the report between, can be easily and objectively measured, and commonly used as basic quality specifications related to maturity of peppers and therefore suitability for harvest. The report total soluble solids (TSS), and titratable acidity (TA) represents an important qualitative parameter TSS/TA. In our investigation the value of this report varied from 21 to 25 (Figure 5).

CONCLUSIONS

For pepper improvement exists sufficient genetic diversity, especially local germplasm can constitute a gene pole that is still insufficient exploited. However, utilization of these resources in breeding programs is time-consuming and resource demanding. To overcome this, pre-breeding activities should be initiated to generate new genetic variability using promising and valuable genetic material for use by the breeders in pepper improvement programs. Results from this experiment provide

evidence that elite pepper materials exist for these characteristics of interest.

This preliminary study can be potentially examined in future investigations and possibly exploited in various breeding methods to maximize their potential superiority as parent material for development of several improved specimens.

Three genotypes present superior qualitative traits: Creola, characterised by 7.90% content in dry substance, TA 0.343 mg g^{-1} malic acid, AA 200.4 mg g^{-1}, and carotenes 23.452 mg g^{-1}; Lider , 7.80 % content in dry substance, TA 0.344 mg g^{-1} malic acid, AA 200.2 mg g^{-1} and carotenes 23.47 mg g^{-1}; Cornel content in dry substance 7.40%, TA 0.331 mg g^{-1}, AA 199.2 mg g^{-1} and carotenes 23.0 mg g^{-1}.

Our study highlight the need of enormous efforts needed to evaluate germplasm for traits of economic importance, for identifying potential donors. The success of pepper improvement program depends on the availability of sufficient genetic variability, but this variability must be in conventionally usable form.

ACKNOWLEDGEMENTS

This research work was cofinanced from PN-II-PT-PCCA-2011-3.2-1351 developed with support of ANCS, CNDI – UEFISCDI, contract registration 68/2012.

REFERENCES

Bubicz M., Perucka I., Materska M., 1999. Content of bioelements of hot. sweet pepper fruit *(Capsicum annuum* L.*)*, Biul Magnezol. 4(2):289-292.
Buczkowska H., Najda A., 2002. A comparison of some chemical compounds in the fruit of sweet, hot pepper *(Capsicum annuum* L.*)*, Fol Hort. 14(2):59-67.
Crosby K.M., Jifon J., Pike L., Yoo K.S., 2007. Breeding vegetables for optimum levels of phytochemicals, Acta Hort. 744:219-224.
Draghici E.M., 2014, Producerea semintelor si a materialului saditor legumicol, Editura Granada, Bucureş ti, ISBN 978-606-8254-41-8.
Herman V.A., 2005. The production of new and improved hot pepper cultivars for the Carribean, *agriculture.gov.bb/files/new%20varieties%20hot%20 pepper%20.pdf,*.
Howard L.R., Talacott S.T., Brenes C.H., Villalon B., 2000. Changes in phytochemical and antioxidant activity of selected pepper cultivars, J. Agric. Food Chem. 48(1):713–720.

Kays S.J., 1991. Postharvest physiology of perishable plant products. Van Nostrand Reinhold, New York, p. 532.

Leskovar D.I., Crosby K., Jifon J.L., 2009. Impact of agronomic practices on phytochemicals and quality of vegetable crops. Acta Hort. 841:317-322.

Lyon B.G., Senter S.D., Payne J.A., 1992. Quality characteristics of oriental persimmons (*Diospyrus kaki*, L.) cv. Fuyu grow in the southeastern United States, J. Food Sci. 57:693-695.

Manela-Azulay M., Mandarim-De-Lacerda C.A., Perez M. de A., Filgueira A.L., Tullia C., 2003Vitamina C, Anais Brasileiro de Dermatologia 78:265-272.

Marcus F., Daood H.G., Kapitany J., Biacs P.A., 1999. Change in the carotenoid and antioxidant content of spice red pepper (paprika) as a function of ripening and some technological factors, J. Agric. Food Chem. 47:100–107.

Navarro J.M., Flores P., Garrido C., Martinez V., 2006. Changes in the contents of antioxidant compounds in pepper fruits at different ripening stages as affected by salinity, Food Chem. 96:66–73.

Pérez-Grajales M., González-Hernández V.A., Peña-Lomelí A., Sahagún-Castellanos J., 2009. Combining Ability and Heterosis for Fruit Yield and Quality in Manzano Hot Pepper (*Capsicum pubescens* R & P), Landraces. Rev. Chapingo Serie Hortic. 15(1):47-55.

Toda Fruta, 2004. O poder de cura das frutas: a fruta campeã de vitamina C, Available in: http://www.todafruta.   com.br/todafruta/ mostra_conteudo, 5571.

6

RESPONSE OF SEXUAL EXPRESSION OF ZUCCHINI SQUASH TO SOME FOLIAR FERTILIZERS TREATMENTS

Dimka HAYTOVA, Nikolina SCHOPOVA

Agricultural University, 12 Mendeleev str., 4000 Plovdiv, Bulgaria
E-mail: haitova@abv.bg

Corresponding author email: haitova@abv.bg

Abstract

The application of foliar sprays is an important crop management strategy, which may help maximizing crop yield and quality. The influence of different agricultural practices, such as foliar application on generative expressions of zucchinis is slightly studied. The objective of our research was to assess the influence of some foliar fertilizers treatments on sexual expression of zucchini squash. The experiments were carried out during the period 2007-2009, on Experimental field of Department of Horticulture at the Agricultural University of Plovdiv, Bulgaria. Variety 'Izobilna F1' was used as an object of the experiments. The field experiments were done by randomized block design with four replications. Complex foliar fertilizers Fitona 3, Hortigrow and Humustimin in three concentrations, separately and in background on soil fertilization $N_{16}P_{16}K_{16}$ were used. The number of male and female flowers per plant, proportion male:female flowers, number of fruits per plant and percentage of fructification were determinate. The results of this experiment indicate that optimal mineral nutrition and providing additional nutrients through foliar application during the period of intensive growth and fruiting influenced positively on the number of fruit formation and increase the percentage of fruit development. The highest number of fruits and the highest percentage of fruit formation is outstanding the variant $N_{16}P_{16}K_{16}$+0.3% Humustim, followed by $N_{16}P_{16}K_{16}$ + 0.2% Hortigrow and by $N_{16}P_{16}K_{16}$ + 0.3% Hortigrow.

Key words: fertilization, foliar application, fruits formation, Cucurbita pepo L.

INTRODUCTION

In monoecious plants, such as cucurbits, the ratio between males and females flowers varies considerably depending on various environmental conditions. The formation and differentiation, as well as their ratio also depend of nutrient regime, on the activity of endogenous phytohormones and by treatment with growth regulators (Lau and Stephenson, 1993; Swiader et al, 1994). Change from vegetative growth to generative stages is a complex process regulated by many factors (Sure et al. 2013).
Studies on specificities of flowering under zucchinis are conducted by many scientists (Nitsch et al, 1952; Yakovlev, 1987; Loy, 2004; Grumet, 2011). One of the main conclusions of the authors is that the formation of female flowers and sufficient male flowers are the limiting factor in production.
The application of foliar sprays is an important crop management strategy, which may help maximizing crop yield and quality (Panayotov, 2004; Panayotov, 2005; Fernandez and Eichert, 2009). The influence of different agricultural practices, such as foliar application on generative expressions of zucchinis is slightly studied.
In this study we aimed to trace the formation of male and female flowers, the ratio between them, and the percentage of initiated fruit set to female flowers in separately foliar application and combining it with soil fertilization.

MATERIALS AND METHODS

The investigations were conducted in the period 2007–2009 under open field conditions with zucchini (*Cucurbita pepo* L. var. *giromontia*), cultivar Izobilna F1 on the experimental field of the Agricultural University of Plovdiv, Bulgaria. The soil of the field is classified as Molic Fluvisols (Popova and Sevov, 2010). The depth of the humus horizon is 28–30 cm. The soil is loamy (clay content from 30% to 41%).
Chemically, the soil is characterized by a low content in organic matter (1.46 %), pH neutral to slightly alkaline (7.17–7.37) and by the

presence of large amounts of $CaCO_3$, which gives more favorable physicalchemical water and soil properties, despite the heavy physical composition. Nitrogen content was low (32–46 mg.kg^{-1}), while there was a good stock of soluble phosphorus (P_2O_5 - 16.7-18 mg.kg^{-1}) and potassium (K_2O – 67 - 96 mg.kg^{-1}).

For the purpose of the experiment three different complete foliar fertilizers were used: Fitona (7.20% N, 5.20% K_2O, 1.5% Ca, 0.9% Mg, 0.1% Fe, 0.1% B, Cu, Zn, Mn, Mo. Fitotech Ltd., Bulgaria), Hortigrow[®] (20% N, 20%, P_2O_5, 20% K2O, 0.06% Fe, 0.02% Zn, 0.01% Mn, 0.01% Cu, 0.02% B, 0.001% Mo and 1% amino acids, Hortiland Ltd,. The Netherlands), Humustim[®] (on base of potassium humates-3% N, 1.14% P_2O_5, 7.83% K_2O, 3.92% Ca, 1.1% Mg, Cu, Zn, Mo, Mn Co, B, S. Agrospeis Ltd., Bulgaria).

Soil fertilization was carried out with NPK using a ratio $N_{160}P_{160}K_{160}$. Phosphorus [Ca $(H_2PO_4)_2$ – 46% P_2O_5]; and potassium (K_2SO_4 – 50% K_2O) fertilizers were applied with last tillage of soil before planting. Nitrogen fertilizer, introduced as NH_4NO_3 (34% N), was applied twice during the growing season.

First application was after formation of new leaves of plants after planting, and the second – 20 days after the first. Water solution of foliar fertilizers was prepared. Foliar fertilizers were applied in the given concentrations three times in the following phases: beginning of flowering, beginning of fruit production and beginning of mass fruit production. Solution with the needed concentration was prepared for the different treatments. Control plants were treated with pure water. The consumption of working solution in the first spraying was 600 l.ha^{-1}, and in the second and third 800 l.ha^{-1}. Plants were cultivated according to the conventional technology for early field production of marrows, using previously produced seedlings (Cholakov, 2009).

The seedlings were planted after thirty days of cultivation in non-heated polythene tunnel. Plants were planted on bed-furrow surface, according to scheme 100+60/50 cm and density of plantation 25000 plants.ha^{-1} in beginning of May.

Growth period was 45 days after planting. Treatments of the experiment:

1. Control - non fertilized;

2. Foliar fertilization with 0.2% Fitona;
3. Foliar fertilization with 0.3% Fitona;
4. Foliar fertilization with 0.4% Fitona;
5. Foliar fertilization with 0.1% Hortigrow;
6. Foliar fertilization with 0.2% Hortigrow;
7. Foliar fertilization with 0.3% Hortigrow;
8. Foliar fertilization with 0.2% Humustim;
9. Foliar fertilization with 0.3% Humustim;
10. Foliar fertilization with 0.4% Humustim;
11. Soil fertilization with $N_{160}P_{160}K_{160}$;
12. Soil fertilization with $N_{160}P_{160}K_{160}$ + 0.2% Fitona;
13. Soil fertilization with $N_{160}P_{160}K_{160}$ + 0.3% Fitona;
14. Soil fertilization with $N_{160}P_{160}K_{160}$ + 0.4% Fitona;
15. Soil fertilization with $N_{160}P_{160}K_{160}$ + 0.1% Hortigrow;
16. Soil fertilization with $N_{160}P_{160}K_{160}$ + 0.2% Hortigrow;
17. Soil fertilization with $N_{160}P_{160}K_{160}$ + 0.3% Hortigrow;
18. Soil fertilization with $N_{160}P_{160}K_{160}$ + 0.2% Humustim;
19. Soil fertilization with $N_{160}P_{160}K_{160}$+ 0.3% Humustim;
20. Soil fertilization with $N_{160}P_{160}K_{160}$ + 0.4% Humustim.

The number of fruits per plant, number of male and female flowers per plant, in tree phases - beginning of fruitfulness, mass fruitfulness and end of fruitfulness were determined. Proportion ♂:♀ was determinate and percentage of fructification.

Statistical analysis: the results were elaborated using the dispersion analysis method for one factor field trial and regression analysis (Dimova and Marinkov, 1999), using the program BIOSTAT (ANOVA).

RESULTS AND DISCUSSIONS

The results of field experiments show that zucchini are consistent in their flowering, despite the known differences in climatic conditions from year to year (Table 1). However, the factors do not change with variations outside the biological requirements of the species, in any of the experimental years. Growing zucchini without soil fertilization and foliar application (control) leads to a lowering of the number of male and female flowers, compared to the other variants of the experiment. At the same time, plants form a smaller number of fruits (Table 1.) and the lowest rate of fructification (Table 2.).

Table 1. Number of fruits per plant, number of male and female flowers per plant, and proportion ♂: ♀, per year and average for 2007-2009

Variants			number of fruits			♂	♀	♂:♀
			2007	2008	2009			
1.	Control		3.50	3.50	3.25	29.75	24.25	1.23
2.	Fitona	0.2%	3.50	4.25	3.75	19.42	16.08	1.21
3.		0.3%	3.50	4.00	3.50	20.17	16.92	1.19
4.		0.4%	3.75	4.00	3.50	22.25	19.08	1.17
5.	Hortigrow	0.1%	3.50	4.25	3.50	22.25	19.00	1.17
6.		0.2%	4.00	4.50	3.75	18.17	16.00	1.14
7.		0.3%	3.25	3.75	3.25	17.08	14.83	1.15
8.	Humustim	0.2%	4.00	4.50	4.25	20.08	18.00	1.12
9.		0.3%	3.75	4.25	3.75	19.25	16.33	1.17
10.		0.4%	3.75	4.00	3.75	15.92	13.67	1.16
11.	$N_{160}P_{160}K_{160}$		4.00	4.50	4.00	14.83	13.00	1.14
12.	$N_{160}P_{160}K_{160}$ Fitona	0.2%	4.00	4.75	4.25	11.75	12.25	0.96
13.		0.3%	4.00	4.50	4.00	12.75	11.58	1.10
14.		0.4%	4.50	5.25	4.50	12.67	11.58	1.09
15.	$N_{160}P_{160}K_{160}$ Hortigrow	0.1%	4.50	5.00	4.75	11.33	11.08	1.02
16.		0.2%	4.50	5.00	4.50	11.00	10.33	1.06
17.		0.3%	5.00	5.50	5.00	10.83	10.92	0.99
18.	$N_{160}P_{160}K_{160}$ Humustim	0.2%	4.75	5.50	4.75	11.75	11.67	1.01
19.		0.3%	5.50	5.75	5.25	10.83	10.67	1.01
20.		0.4%	5.25	5.75	5.00	13.75	12.58	1.09

This specificity is most likely due to the fact that the plants are placed in conditions of lack of nutrients to ensure that the vegetative growth, normal course of flowering and fruit set of formed. Additionally, the reason can be found in the large number of aborted flowers, which is most likely due to the poor quality of pollen fertility declining ability under the influence of foliar fertilization applied alone or with soil fertilization there are changes in the number of flowers of each gender, by year and average for the period.

The number of male flowers on average for the period decreased by 29.75 units for the control to 10.83 units for $N_{160}P_{160}K_{160}$ + 0.3% Humustim and $N_{160}P_{160}K_{160}$ + 0.3%Hortigrow.

The number of female flowers remains relatively constant at all tested variants both in years and average for the period. In variants of separately foliar application this number is slightly higher than those in which plants are grown $N_{160}P_{160}K_{160}$ background. On the other hand, in these variants are formed by a smaller number of fruits (Table 1.) and fructification rate is lower (Table 2).

Table 2. Percentage of fructification in different variants of fertilization

Variants			Average for year		
			2007	2008	2009
1.	Control		15.24	14.05	13.21
2.	Fitona	0.2%	23.41	25.66	22.35
3.		0.3%	21.93	22.55	20.17
4.		0.4%	21.42	20.13	17.78
5.	Hortigrow	0.1%	19.49	21.66	17.78
6.		0.2%	26.88	27.78	23.27
7.		0.3%	21.94	24.22	21.75
8.	Humustim	0.2%	23.26	23.85	21.61
9.		0.3%	23.33	25.50	22.12
10.		0.4%	30.28	28.37	29.12
11.	$N_{160}P_{160}K_{160}$		32.32	33.67	30.44
12.	$N_{160}P_{160}K_{160}$ Fitona	0.2%	36.91	35.89	35.04
13.		0.3%	36.91	36.99	37.11
14.		0.4%	40.65	43.85	38.33
15.	$N_{160}P_{160}K_{160}$ Hortigrow	0.1%	42.59	47.06	41.61
16.		0.2%	45.09	48.61	44.00
17.		0.3%	50.30	50.13	42.61
18.	$N_{160}P_{160}K_{160}$ Humustim	0.2%	44.04	45.79	40.09
19.		0.3%	53.81	50.73	47.81
20.		0.4%	45.45	45.05	40.01

Zucchinis as annual crops for their short growing season are unable to keep plenty of underlying flowers, as well as regulating and preserving formed fruits. The number of male flowers is reducing. The ratio ♂: ♀ is changing, respectively of 1.22 for the control, to 1.00 for variants which utilize Humustim at concentrations 0.2% and 0.3% and fertilization with $N_{16}P_{16}K_{16}$. In 2008 and 2009 the ratio ♂: ♀ is slightly lower than 1.00 only for $N_{16}P_{16}K_{16}$ + 0.2% Fitona, respectively 0.98 and 0.96. The result is probably an anomaly caused by the fluctuation in the number of male flowers.

Used for the experiment cv. Izobilna F1 is characterized by continuous flowering in terms of early field production. Differences between variants are insignificant. Not observed deviations from normal course of flowering, which are caused as a result of the use of fertilizers. Foliar fertilizers applied separately or soil fertilization background with $N_{160}P_{160}K_{160}$ affects the number of initiated fruits (Table 1). The amendments between variants are small, but with a greater number of shaped fruit per plant are distinguished those with mixed fertilization (soil and foliar). Fertilization $N_{16}P_{16}K_{16}$ + 0.3%Humustim causes the formation of the largest number of fruits per plant in comparison with the control. The same trends were observed in the percentage of fructification (Table 2). The highest percentage of fructification has plants fertilized with $N_{160}P_{160}K_{160}$ + 0.3%Humustim in the three years of experiments, respectively, 53.81%, 50.73% and 47.81%. Immediately after them of rank are $N_{160}P_{160}K_{160}$ + 0.3% Hortigrow, $N_{160}P_{160}K_{160}$ + 0.2% Hortigrow and $N_{160}P_{160}K_{160}$ + 0.4% Humustim. Adequate mineral nutrition and additional nutrients in an easily absorbable form in periods of rapid growth and fructification have a positive impact on the number of fruits and increase the percentage of fructification. The reason for such a reaction of zucchinis, Stephenson et al. (1988) found in "dominance of the first fruits", which temporarily inhibit flowering during the growth of fruits.

CONCLUSIONS

The results of the assays on the influence of foliar fertilization on the biological behaviours of zucchini indicated that the use of foliar fertilizers during the growing season have a positive influence on the growth and development of plants.

Optimal mineral nutrition and providing additional nutrients through foliar application during the period of intensive growth and fruiting influenced positively on the number of fruit formation and increase the percentage of fruit development.

The highest number of fruits and the highest percentage of fruit formation is outstanding the variant $N_{160}P_{160}K_{160}$+0.3% Humustim, followed by $N_{16}P_{16}K_{16}$ + 0.2% Hortigrow and by $N_{160}P_{160}K_{160}$ + 0.3% Hortigrow.

ACKNOWLEDGEMENTS

The authors would like to thank the Research Fund of the Agricultural University of Plovdiv, Bulgaria for financial support.

REFERENCES

Cholakov D. 2009: Technology for cultivation of marrows in vegetable-growing, Academic publishers of Agricultural University -Plovdiv, pp. 150-158 (in Bulgarian).

Dimova D., Marinkov E. 1999. Experimental work and biometry, Academic publishers of Agricultural University, Plovdiv (in Bulgarian)

Fernández V, Eichert T .2009. Uptake of hydrophilic solutes through plant leaves: current state of knowledge and perspectives of foliar fertilization. Crit Rev Plant Sci 28:36–68

Grumet R., Taft J. 2011. Sex expression in cucurbits, Genetics, Genomics and Breeding of Cucurbits, p. 353-375

Lau, T. C.; Stephenson, A. G., 1993. Effects of soil nitrogen on pollen production, pollen grain size, and pollen performance in Cucurbita pepo (Cucurbitaceae)., American Journal of Botany 80 (7) : 763-768

Loy J.B. 2004. Morpho-Physiological aspects of productivity and quality in squash and pumpkins (Cucurbita spp.), Critical Reviews in Plant Sciences, 23(4):337-363

Nitsch J.P., Kurtz E.B., Liverman-Jr J.L., Went F.W. 1952. The development of sex expression in cucurbit flowers, American Journal of botany, vol. 39, №1, p.32

Panayotov, N. 2005 Morphological behaviours and productivity of pepper plants under influence of foliar fertilizer Kristalon. Analele Universitii "Valahia" Targoviste. Facilia VI "Tehnologia produselor alimentare, pescuit si acvacultura". The annals of "Valahia" University of Targoviste, Fascicle VI "Food technology, aquaculture and fishing", pp, 24-30.

Panayotov, N., 2004. Morphological development and productivity of pepper plants after application of foliar fertilizer Hortigow. Scientific researches on the Union of Scientists in Bulgaria-Plovdiv, Series B "Technique and Technology," Scientific session "Technology, Agricultural Sciences and Technology" –vol.III, pp 97-104 (in Bulgarian).

Popova.R, Sevov A., 2010. Soil characteristic of experimental field of crop production department as result of cultivation of grain. technical and forage crops. Agricultural University of Plovdiv, Scientific works, vol.LV, book1, pp. 151-156 (in Bulgarian).

Stephenson, A.G., Devlin, B., Horton, J.B. 1998. The effect of seed number and prior fruit dominance on the pattern of fruit production in Cucurbia pepo(zucchini squash). Annals of Botany 62:653-661.

Sure S., Arooie H, Azizi M. 2013. Effect of GA3 and Ehtephon on sex expression and oil yield in medicinal pumpkin (Cucurbita peop var. styriaca). IJFAS Journal-2013-2-9/196-201.

Swiader J.M., Sipp S.K., Brown R.E. 1994. Pumpkin growth, flowering and fruiting response to nitrogen and potassium sprinkler fertigation in sandy soil, J. Amer. Soc. Hort. Sci. 199(3):414-419

Yakovlev, MS 1987. Embryology of cereals, legumes, vegetables cultivated plants. Kishinew, "Shtiintsa. Russian.

7

RESEARCH REGARDING THE EFFECT OF APPLYING HERBICIDES TO COMBAT WEEDS IN QUICKLY-POTATO CULTURES

Gheorghița HOZA[1], Bogdan Gheorghe ENESCU[1], Alexandra BECHERESCU[2]

[1]University of Agronomic Sciences and Veterinary Medicine of Bucharest, 59 Mărăşti Blvd, District 1, 011464, Bucharest, Romania, Email: hozagh@yahoo.com

[2]Banat University of Agricultural Sciences and Veterinary Medicine Timisoara, Calea Aradului nr. 119, 300645 Timişoara, Jud. Timiş, România,
Email: alexandra_becherescu@yahoo.com

Corresponding author email: hozagh@yahoo.com

Abstract

The infields from our country have a very high degree of becoming full of weeds because its composition, the ratio of participation of various species and the degree of infestation vary and modify according to the region, vegetable crop rotation, applied combat measures and technology. The research was conducted in the vegetable region of Lunguletu, Dambovita county, with quickly-potato, 'Impala' variety, applying the herbicides Sencor 70 WG 1 kg/ha and Titus 25 DF 50g/ha, applied pre-emergent, post-emergent and combined, according to the action manner of the products, in order to establish the effectiveness of weed combat and level of damages caused by weeds. It was noticed that the best results were obtained by applying both herbicides, the degree of weed combat being 91%, compared to applying only one product, where the combat degree was 57% for Sencor 70 WG and 78 % for Titus 25 DF. The tubercle production was influenced by the degree of weed combat, the highest production being obtained for the variant that recorded the highest degree of combat, meaning the variant with both products applied.

Key words: potato, herbicides, production

INTRODUCTION

Potato represents the second species within human diet after the basic cereals, because it is a complex aliment, substantial, rich in vitamins and minerals. Obtaining large potato productions involves practicing the specific culture technology, in compliance with all technological phases. Among these, a special role is played by combating weeds, which can cause serious damages by competing with the culture plants for water, air, light and nutrients. Weeds have a high reproduction capacity; their seeds have a wide lifespan and the capacity to germinate gradually for a long time, thus being able to compromise the culture when controlling them may result difficult.

In case of potatoes, after the plants have sprung up and covered the soil and the effects of herbicides applied pre-emergently, late weeds appear, which influence less the production but hinders the mechanical harvesting, increases the percentage of mechanical damaging of the tubercles and of losses, reduces the productivity of the harvesting machines (Berindei, M. 1985; Frâncu, Georgeta, 1995; 1996; 1998,; 1999). Weeds can represent bridges for the diseases and pests to be transmitted from one year to the next or from one culture to another oen, because many species of weeds represent hosts for different pests (Frâncu, Georgeta 1996 b; Ianosi, S. şi Boţoman, Gh. 2004).

The most dangerous weed, which can cause serious problems for combating, reproduce very fast, invade the potato culture and threaten the production, are called *"problem weeds"*. The majority of these are perennial species that reproduce both by seeds and by vegetative reproduction, whose number and diversity across large areas increased with the unilateral use of herbicides that these species are resistant to (Berindei, M. 1985; Frâncu, Georgeta, 1987; Ianosi, S. 2002).

MATERIALS AND METHODS

The experiment was organized in Lunguleţu, Dâmboviţa County. The quickly-potato culture was realized on an non-evolved alluvial soil, with loamy-sandy texture, the thickness of horizon A of 21 cm, content in humus of 1.24% (weakly ensured), content in clay of 1,02%, mobile phosphor of 21.4 mg/100 g soil, mobile potassium of 60 mg/100 g soil and pH value of 6.2.

The biological material was represented by the IMPALA soil, with a vegetation period of 80-90 days, resistant to nematodes (Ro$_{1-4}$) and a medium content of starch of 14 %. For weed combat, two herbicides were used: Titus 25 DF and Sencor 70 WG, applied separately and combined, according to the experimental scheme:

V1 = control 1 not worked and without herbicides.

V2 = control 2 worked both mechanically and manually and without herbicides.

V3 = SENCOR 70 WG-1,0 kg/ha applied (pre-emergent).

V4 = SENCOR 70 WG-1,0 kg/ha (pre-emergent)+TITUS 25 DF- 50 g/ha applied (post-emergent).

V5 = TITUS 25DF- 50 g/ha applied (post-emergent).

Planting density was 57.000 nests/ha (70 x 25 cm), and the density after springing up was 51.000 nests/ha. There was an agricultural basis of: N150, P150, K150 kg/ha ensured by applying 1000 kg of the chemical fertilizer Complex 15:15:15 when preparing the soil. The pre-emergent culture was autumn cabbage.

Working method, observations and measurements

a.*Mapping weeds*

The measurement of weeds was made by using a metric frame, recording the number of weeds per square meter. The result from the control 1 (V1) represents the first level of infestation, while the results from the variants V3-V5 represent the combat effects of the used herbicides.

In order to evaluate the degree of weed infestation and effectiveness of control methods, the following evaluation methods were used:

- covering degree *(G. a. %) = (no. of weeds per variant / no. of weeds for control 1) x 100.*

- combat degree *(G. c. %) = 100 - G. a.*

- participation degree (G. p. %) = (no. of weeds per species / total no. of weeds per variant) x 100.

b. *Production analysis*

At the final harvest (20 June) the total production, the consumption fraction (tubercles over 30 mm diameter) and the under STAS fraction (tubercles under 30 mm diameter) were measured.

The estimation of damages produced to the culture by weeds was made based on the mathematical relations used in plant protection, proposed by Rotaru, V; Mihăiță, A; Alexandri, Al. (1999), by using the following formulae:

$P = (1 - q_0 / q_1) x 100$ where:

P = damage (%),

q_0 = average production for the variant with weeds (t/ha),

q_1 = average production for the variant without weeds – control 2 (t/ha)

RESULTS AND DISCUSSIONS

a. *Number of weeds measured for the control 1 – not worked and without herbicides*

The number of per species and per group (plants/m^2), as well as the participation ratio (%) to the infestation degree for the control 1 variant is different depending on the weed species. Among annual weeds, the largest numbers of plants for one species were recorded for bristle grass (*Setaria glauca*) 31 plants/m^2, for amaranth (*Amaranthus retroflexus*) 12 plants/m^2; for orache (*Chenopodium album*) 11 plants/m^2 (Table 1).

Among the perennial dicotyledonous species, the following numbers were recorded: 6 plants/m^2 for pelamid (*Cirsium arvense*); 5 plants/m^2 for bindweed (*Convolvulus arvensis*); 5 plants/m^2 for sow thistle (*Sonchus arvensis*); for the perennial monocotyledonous the largest number was recorded for cane (*Sorghum halepense*) 4 plants/m^2.

Among the species with large number of plants, the bristle grass (*Setaria glauca*) participates to the infestation degree by an average of 34 % the amaranth (*Amaranthus retroflexus*) by 13 %, the orache(*Chenopodium album*) by 12 %, the pelamid (*Cirsium arvense*) by 8 % and the bindweed (*Convolvulus arvensis*) by 5%.

b. *Number of weeds per species for the variants with herbicides*

The herbicides used (V3-V5) totally combated (table 2) the annual monocotyledonous weeds. The perennial monocotyledonous weeds, represented by cane (*Sorghum halepense*) were combated for the variant with SENCOR 70WG 1 kg/ha applied pre-emergently and TITUS 25DF 50 g/ha applied post-emergently (V4), but remained partially combated for the variants with only one herbicide SENCOR 70WG 1 kg/ha applied pre-emergently (V3) and TITUS 25DF 50 g/ha post-emergently (V5). Regarding the combating of dicotyledonous weeds, it could be observed that, except for the V3 were weeds were not combated, for the majority of variants the results were good. The major problem is represented by *Convolvulus arvensis* (bindweed) that was not combated in any of the variants. However, it could be observed that for the variant V4 the rest of the weeds were successfully combated by applying herbicides both pre-emergence and post-emergence.

perennial dicotyledonous weeds were not combated. As a result of applying this herbicide, the covering degree was 43% and the combat degree was 57%.

For V4, the combination between the two herbicides, SENCOR 70WG and TITUS 25DF, ensured a culture with very few weeds, resulting a covering degree of 9% and a combat degree of 91%.

For V5, the variant only with post-emergence TITUS 25DF, the results were good, the herbicide having proper effects both on monocotyledonous and dicotyledonous weeds. For this variant, with 20 plants/m^2 not combated, the covering degree was 22% and the combat degree was 78%.

Table 1. Number of weeds measured for the control 1 – not worked and without herbicides

Group of plants/species	Nr./m^2	G. p. %
Annual monocotyledonous		
Setaria glauca	31	34
Total annual monocotyledonous	*31*	*34*
Perennial monocotyledonous		
Sorghum halepense	4	4
Total perennial monocotyledonous	*4*	*4*
Total monocotyledonous	*35*	*38*
Annual dicotyledonous		
Amaranthus retroflexus	12	13
Brassica rapa	7	8
Chenopodium album	11	12
Galinsoga parviflora	7	8
Polygonum persicaria	3	3
Total annual dicotyledonous	*40,0*	*44*
Perennial dicotyledonous		
Cirsium arvense	6	8
Convolvulus arvensis	5	5
Sonchus arvensis	5	5
Total perennial dicotyledonous	*16*	*18*
Total dicotyledonous	*56*	*62*
TOTAL weeds (nr./m^2)	91	100

Table 2. Number of weeds per species for the variants with herbicides

Group of plants/species	V1 Mt.1	V 3	V 4	V 5
Annual monocotyledonous				
Setaria glauca	31	0	0	0
Total annual monocotyledonous	*31*	*0*	*0*	*0*
Perennial monocotyledonous				
Sorghum halepense	4	3	0	2
Total perennial monocotyledonous	*4*	*3*	*0*	*2*
Total monocotyledonous	*35*	*3*	*0*	*2*
Annual dicotyledonous				
Amaranthus retroflexus	12	10	3	1
Brassica rapa	7	0	0	2
Chenopodium album	11	8	0	2
Galinsoga parviflora	7	0	0	0
Hibiscus trionum	-	-	-	-
Polygonium persicaria	3	2	0	0
Sonchus oleraceus	-	-	-	-
Total annual dicotyledonous	*40*	*20*	*3*	*5*
Perennial dicotyledonous				
Cirsium arvense	6	6	0	2
Convolvulus arvensis	5	5	5	5
Sonchus arvensis	5	5	0	2
Total perennial dicotyledonous	*16*	*16*	*5*	*9*
Total dicotyledonous	*56*	*36*	*8*	*18*
TOTAL weeds	91	39	8	20
covering degree *G.a. %)*	100	43	9	22
combat degree (*G.c. (%)*	*0*	*57*	*91*	*78*

For V3, the pre-emergent application of SENCOR 70WG led to the combat of annual monocotyledonous weeds (*Setaria glauca*) and of several annual dicotyledonous weeds. The

The economic analysis of the efficiency of weed combat in quickly-potato culture was performed according to the obtained productions and their value for the studied variants. These results are presented in figure 1, where it can be noted that for the control variant 2 (V2-worked manually and mechanically), which recorded the highest cost (500 lei/ha), the combat degree was rather high (91 %).

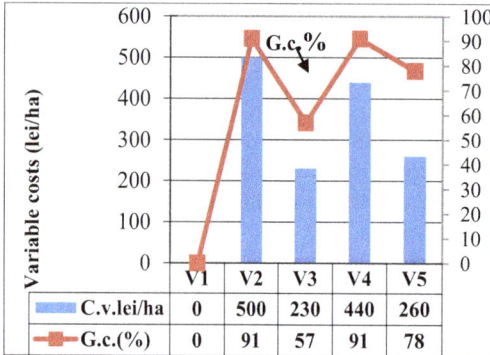

Figure 1. Treatment cost and combat degrees per variant

The variant with pre-emergence herbicide recorded the lowest costs, meaning V3 only with SENCOR 70Wg 1 kg/ha applied, the cost level was 230 lei/ha and the combat degree was 57 %. For V4, where both herbicides were applied, SENCOR 70WG - 1 kg/ha and TITUS 25DF – 50g/ha, the cost was 440 lei/ha, but with a combat degree of 91 %. For the variant only with TITUS 25DF -50 g/ha applied (V5), the cost was 260 lei/ha for a combat degree of 78%. The total tubercle production, per consumption fraction and under STAS fraction (fig. 2) and the percentage allocation (table 3) was influenced by the application of herbicides.

Table 3. Percentage allocation of the production of tubercles for consumption and under STAS from the total production

Variant	% of production	
	consumption	under STAS
V1 (control 1 not worked)	89.8	10.2
V2 (control 2 worked)	95.0	5.0
V3 (Sencor 70WG)	91.8	8.2
V4 (Sencor 70WG+Titus 25DF)	92.5	7.5
V5 (Titus 25DF)	91.4	8.6
Average	92.1	7.9

Regarding the total production, it was noted that for the variant where pre-emergent SENCOR was used, the obtained production

was of 28.2 t/ha, while for the variant where post-emergent TITUS 25DF was used the production obtained was 30.3 t/ha. For the V4 variant, where two herbicides were applied, the production was 36 t/ha. The additional post-emergent herbicide with TITUS 25DF influenced the total tubercle production due to the wider span of weeds combated by that particular herbicide.

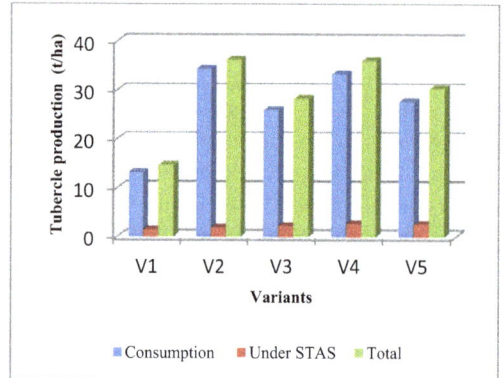

Figure 2. Tubercle production, total and per fractions (t/ha)

The production of tubercle for consumption was only of 13.2 t/ha for the control variant 1 (V1), while for the control variant 2 (V2) the production was of 34.4 t/ha; 21.2 t/ha larger due to combating the weeds and soil loosening. As in the case of total production, the production of tubercles for consumption for the control variant 1 (13.2 t/ha) was lower than for the other variants, while the production for the control variant 2 (34.4 t/ha) was higher than for the variants with herbicides, except for the one with both herbicides - SENCOR 70WG and TITUS 25DF (V4).

For the variant with SENCOR 70WG, the production of tubercles for consumption was of 25.9 t/ha, for the variant with TITUS 25DF the production was of 27.7 t/ha, and for the variant with both of them (V4) the production was of 33.3 t/ha.

The production of tubercles under STAS (with diameter under 30 mm) generally represent 10-15 % from the total tubercle production; in this case it represents 7.8 %.

The damages caused by weeds to the total tubercle production (table 4) are serious for the variants without combat measures, 59 % for

V1, compared to 0,5 % for the combined application of the herbicides Sencor and Titus

and 0 % for soil maintenance.

Table 4. Damage caused by weeds to total production and to the production for consumption

Variant	Total production		Production for consumption	
	Prod. t/ha	Damage %	Prod. t/ha	Damage %
V1 (control 1 not worked)	14.7	59	13.2	61
V2 (control 2 worked)	36.2	0	34.4	0
V3 (Sencor 70WG 1 kg/ha)	28.2	22	25.9	25
V4 (Sencor + Titus)	36.0	0,5	33.3	3
V5 (Titus 25DF 50g/ha)	30.3	16	27.7	19
Average per variant	29.08	24.3	26.9	27.0

CONCLUSIONS

The highest level of weeds was produces by the annual dicotyledonous and monocotyledonous species (80 %), which must be taken into consideration when choosing herbicides, because these species can be combated relatively easy and cheaper by applying herbicides pre-emergently

The analysis of weed combat effectiveness, for the herbicides applied pre-emergently to the quickly-potato from Lunguleţu, showed that for the variant with pre-emergent application of the herbicide SENCOR 70WG the combat degree was 57 %, for the variant with post-emergent application of TITUS 25DF the degree was 78 %, and for the double herbicide variant, with pre-emergent SENCOR and post-emergent TITUS, the combat degree was 91%

The economic study related to costs and combat degree conducted on the variants showed that for the control 2 (V2 worked), for which he highest cost of 500 lei/ha was recorded, the combat degree was 91 %. The variant with only pre-emergent herbicide recorded the lowest cost, thus for SENCOR 70WG a cost of 230 lei/ha and combat degree of 57 %, while for the variant with only post-emergent the cost was 260 lei/ha and the combat degree of 78%. The pre-emergent application of SENCOR 70WG and post-emergent application of TITUS 25DF, led to a cost of 440 lei/ha and a combat degree of 91%, the same result as for variant 2.

Among the weed combat variants for the quickly-potato cultivated in Lunguleţu, the lowest productions were obtained from control 1, while the highest were obtained from the control 2 and the variants with herbicides. The productions obtained from variant 2 or the

variant with both herbicides were significantly higher than the other variants with herbicides applied. The degree of weed infestation strongly influenced the level of tubercle production.

REFERENCES

Berindei, M.,1985. *Combaterea integrată a buruienilor din culturile de cartof.* În, Cultura cartofului (Ghidul fermierului); Edit. Ceres, Bucureşti.

Berindei, M., 1999, *Cartoful pentru consum timpuriu trebuie şi poate fi cultivat în toată ţara.* În, Rev. Cartoful în România, Braşov.

Frâncu, Georgeta.1995. *Aplicarea postemergentă a erbicidelor la cultura carto-fului.* În, Rev. Cartoful în România, Braşov

Frâncu, Georgeta,1996. *Combaterea buruienilor din cultura cartofului,* În, Ghid practic de protecţie a cartofului, Edit. Ceres, Bucureşti

Frâncu, Georgeta, 1998. *Aplicarea preemergentă a erbicidelor la cultura cartofului.* În, Rev. Cartoful în România, Braşov

Ianosi, S., 2002. *Lucrările de întreţinere, înainte şi după răsărirea cartofului.* Cultura cartofului pentru consum; Edit. Phoenix, Braşov

Ianosi, S.; Boţoman, Gh., 2004. *Pagube produse de buruieni în cultura carto-fului.* În, Combaterea buruienilor din culturile de cartof. Edit. Phoenix, Braşov.

8

INFLUENCE OF ILLUMINATION WITH LEDs ON GROWTH AND DEVELOPMENT OF LETTUCE SEEDLINGS

Elena PANȚER[1], Maria PELE[1], Elena Maria DRĂGHICI[2]

[1]University of Agronomic Sciences and Veterinary Medicine of Bucharest, Faculty of Biotechnologies, 59 Mărăşti Blvd, District 1, 011464, Bucharest, Romania
Email: mpele50@yahoo.com

[2]University of Agronomic Sciences and Veterinary Medicine of Bucharest – Faculty of Horticulture, 59 Mărăşti Blvd, District 1, 011464, Bucharest, Romania,
Email: elena.draghici@horticultura-bucuresti.ro
Corresponding author email: elena.draghici@horticultura-bucuresti.ro

Abstract

The paper aimed to present the evolution of lettuce (Lactuca sativa L.) grown in different light conditions. Plants were grown in a climate chamber of the Faculty of Horticulture-UASVM Bucharest, using as light sources LED and neon. There have used two varieties of lettuce namely 'Lollo Rosa' and 'Lollo Bionda'. Both of varieties have better development (eg germination period, number of leaves) in the case of LED luminaries compared with those developed under neon light. Thus, masses obtained after 30 days of growth were 30.8% higher in case of the variety 'Lollo Rosa' and 29.8 for variety 'Lollo Bionda'. As a general conclusion it can be say that use of LED light has a lot of advantages against neon light being ecological and offering a reduced energy consumption which determine a lower greenhouse warming phenomenon compared to overheating caused by neon lighting. In addition, it is important that to pursue the studies by assessing the behavior of plants at different radiation emitted by the LED.

Key words: lettuce, LED illumination, growth characteristics.

INTRODUCTION

Lettuce (*Lactuca sativa* L.) is a species commonly cultivated in protected areas especially during the autumn-winter-spring.
For some varieties of lettuce external factors such as light and temperature can influence the quality leaves or heads. For example Piroga lettuce varieties (var. Crispa) and Paris White (var. Romanian) under short-day conditions with increased cloudiness and low temperature (2-4°C) leaves occurring embossed, coarser, crunchy, stained, without commercial aspect. To these varieties leaf tenderness may return only when the temperature rises above 7°C and light conditions are improved (Draghici, 2014). Light is one of the most important environmental factors that exerts different influences on plants, it is not only a source of energy in photosynthesis, but also the promoter of germination, growth and development of plant organisms. The reaction of plants to light, counts both the quantity and quality of radiation incidents as well as length of the lighting period, respectively photoperiod.

As stated Tikhomirov (1994), spectral changes of light determines the recording - to the green organs of each plant – of morphogenetic and different photosynthetic reactions with intra- and interspecific variations or even intra-individuals.
LEDs are widely used in various fields of activity. In the last years, there were indoor plant growers who use LED light to grow plants used for their own food (Yeh and Chung, 2009). Light-emitting diodes (LEDs) have long been recognized for their benefits, long life, energy efficiency and flexibility.
In the last decade LED technology evolution has been outstanding, the market being rising significantly in recent years. However, in comparison with the rest of the lighting industry, the introduction of LEDs in horticultural cultivation techniques is more recent being materialized only in recent years. Introducing of LEDs on the horticultural market in is largely due to significant advances in technology (Matioc, 2012).
In recent years, studies have shown that plants are more sensitive to certain wavelengths of

light to increase the absorption of chlorophyll and photosynthesis standing out exposure to red (~ 640-660 nm) and blue (~ 450 nm) light. These studies showed that the peaks of the spectral light absorption for the *chlorophyll a* is within the ranges of 400 to 500 nm and 600 to 700 nm. The use of additional lighting using lamps helps to produce greenhouse vegetable salad and their use has skyrocketed over the past 15 years in the Netherlands. The main reasons for using additional lighting are ensuring high yields throughout the year and the level of quality that meets market demand (Marcelis et al., 2002).

In the Netherlands, in 1999 additional illuminated surfaces in greenhouses increased by about 13% compared to 1994 (Bakker et al., 2000). Until 2002 about 22% of the Dutch greenhouse made use of supplementary lighting, the digit having increased by 1.7 percent annually compared to 1994 (Knijff and Benning, 2003).

The study aimed to identify the best light spectrum to produce seedlings of lettuce.

MATERIALS AND METHODS

The experiment was conducted in the Department of Bioengineering Hortiviticultural Systems, Faculty of Horticulture Bucharest, during the period 2013-2014.

The study was conducted in the climate chamber, under controlled environment, ensuring a temperature of 22 °C day and 20 °C night, a constant atmospheric humidity of 65%. The duration of lighting was 16 hours / day light and 8 hours dark.

Have been sown seeds of lettuce varieties Lollo Bionda and Lollo Rosa, 150 seeds for each varieties, in three repetitions.

Determinations were carried out on emergence, the growth of salad plants, dynamics of leaves formation, the seedlings mass. It used one type of LEDs.

RESULTS AND DISCUSSIONS

It was found that after three days from sowing that the highest percentage of emergence was recorded at Lollo Bionda variety (20%), the variant exposed to light LED.

At the all variants, the emergence of plants was higher for all cultivars exposed to LED light (figure 1).

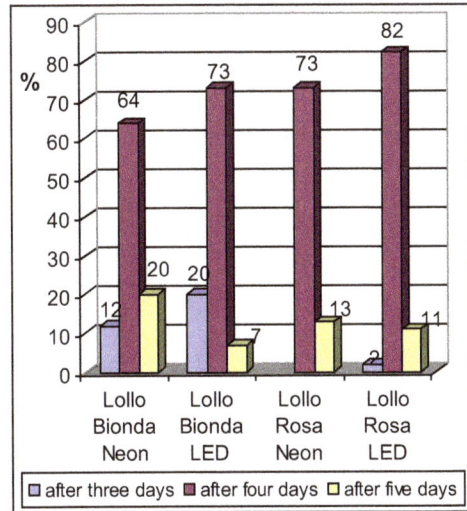

Figure 1. The duration and the percentage of the plants emergence.

Notice that the highest percentage of emergence was Lollo Bionda variety of 100% followed by 95% variety Lollo Rosa to variants lighting by LED, figure 2.

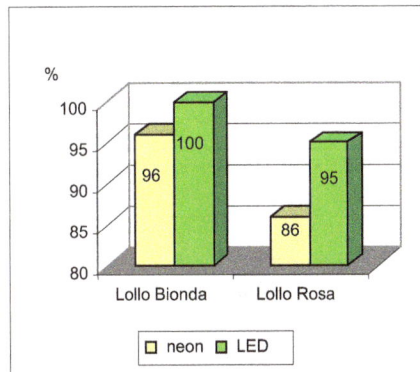

Figure 2. The total percentage of seeds emerged

Observations on the dynamics of lettuce leaves, Lollo Bionda and Lollo Rosa varieties, showed that for both varieties has been formed a greater number of leaves when seedlings growth under LED lighting (figure 3 and figure 4).

Likewise, following the plant masses after 30 days it can see that these have the same aspect. Namely, the salad masses after 30 days of development is higher at both varieties by about 30% (30.8% Lollo Rosa and 29.8% for

Lollo Bionda) to plants grown under illumination provided by LED compared to those grown under neon lighting.

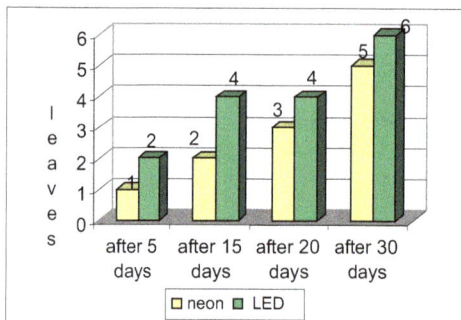

Figure 3. The dynamics of formation of leaves for the variety Lollo Bionda.

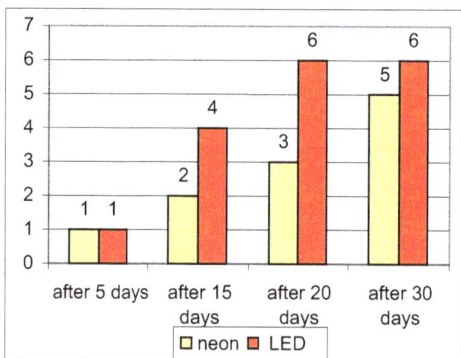

Figure 4. The dynamics of formation of leaves for the variety Lollo Rosa

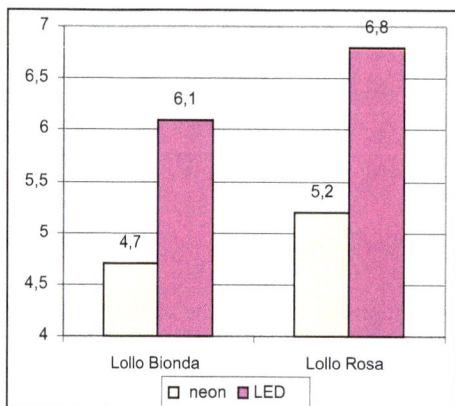

Figure 5. The mass of seedlings after 30 days

CONCLUSIONS

The results obtained clearly confirmed by the fact that they were very close to rehearsals conducted under the same conditions (3 repetitions for the same conditions) shows the obvious advantage of using LED luminaries.
Important is that they are ecological and of reduced energy consumption point of view which determine a lower greenhouse warming phenomenon compared to overheating caused by neon lighting.

ACKNOWLEDGEMENTS

This work was supported by a grant of the Romanian National Authority for Scientific Research, CNDI – UEFISCDI, financed from project number PN-II-PT-PCCA-2011-3.2-1351 - Contract No.68/2012 and the European Social Found, Human Resources Development Operational Programme 2007-2013, project no. POSDRU/159/1.5/S/13276.

REFERENCES

Drăghici Elena Maria, 2009, Producerea semințelor și materialului săditor legumicol, Editura Atlass Press, București.
Knijff și Belnninga, 2003, Climate change, elevational range shifts, and bird extinctions. *Conservation Biology, 22*, 140–150.
Matioc Mirela-Maria, 2012, Studiul influenței luminii produse de LED-uri (light-emitting diodes) asupra germinației și creșterii plantulelor, Teză de doctorat
Marcelis et all., 2002, Growing Plants Under Artificial Lighting [in Russian], Leningrad.
Tikhomirov, A.A. 1994. Spectral composition of light and growing of plants in controlled environments, p 25-29. In: T.W.Tibbitts (ed.). International Lighting in Controlled Environments Workshop, NASA-CP-95-3309.
Yeh and Chung, 2009, Director Soil and Plant Laboratory, Inc. Perlite vs. Polystyrene in Potting Mixes, Santa, Ana, California

9

THE REACTION OF TOMATO GENOTYPES TO FUNGAL PATHOGENS UNDER CONTROLLED CONDITIONS

Nadejda MIHNEA, Galina LUPA ŞCU, Irina ZAMORZAEVA

Academy of Sciences of Moldova, Institute of Genetics, Physiology and Plant Protection, 20 Pădurii Street, MD-2002, Chisinau, Republic of Moldova

Corresponding author email: mihneanadea@yahoo.com

Abstract

The paper presents results of the reaction of new tomato varieties and perspective lines created in the Institute of Genetics, Physiology and Plant Protection to the seeds treatment with FC Fusarium oxysporum, F. solani, F. redolens, Alternaria alternata based on the estimation of seeds germination and length of embryonic radicles and stemlets. It should be mentioned that the reaction of some varieties to same isolates was quite different. Thus, in the case of seeds germination the capability of some approved varieties after treating them with FC of fungi, the highest resistance was found for varieties 'Exclusiv', 'Desteptarea', 'Milenium'. Variety 'Mihaela' has demonstrated the highest complex resistance to FC of 4 used fungi in the case of the reaction of radicles and stemlets, the phenomenon was manifested as insignificant reduction or stimulation. Using bifactorial analysis of variance we found that the most contribution to the variation of embryonic radicle and stemlet length had the species of fungus, being 41.07 and 58.58%, respectively. It should be mentioned that the genotype plays an important role too, its factorial pondering consists of the 32.49 and 22.00%, respectively, for the length of radicles and stemlets, and the interaction tomato genotype x fungus species was recorded as 25.03 and 17.73%, respectively, for both characters. The significant pondering of the influence of fungus species (Fusarium spp., A. alternata) and the interaction tomato genotype x species of fungus on the source of variation of plant growth organs reveals the necessity of constant monitoring of the composition of pathogen species and their virulence.

Key words: tomato, resistance, fungal pathogens, Fusarium spp., A. alternata.

INTRODUCTION

The resistance of agricultural crops to biotic and abiotic stressful factors is one of the basic requirements to contemporary varieties, including varieties of tomato. The problem of complex and sustainable resistance is real for many crops, but especially for tomato because their fruits are extensively used in ordinary food or dietary management of children and elderly people that is why the application of chemical plant protection has to be limited. In the Republic of Belarus fungal diseases may cause losses of 40-60% of tomato crops, in some years even of 80% (Polixenova, 2009).

Treatment with fungicides is quite difficult to optimize or make more efficient because as the capacity of plant infection as well as the efficiency of preparations may be manifested only in favorable climatic conditions that cannot often be forecasted.

The diseases are easily transmitted from one plant to another, spreading extensively on large surface. Tomatoes are susceptible to 200 fungal, bacterial, viral diseases and nematodes. The most common diseases are root rot, *Fusarium* wilt, *Verticillium* wilt, blight, various bacterial and viral infections, for which some sources of resistance are identified (Foolad, 2007; Scott Gardner, 2007).

Toxins of fungi *Alternaria* spp. and *Fusarium* spp. are often involved in the pathogenesis, accumulating in tomato fruits, especially of susceptible varieties, and making them quite toxic for human (Yamagishi et al., 2006).

Performing agricultural techniques, biological control, resistant varieties, chemical treatments are considered as basic factors of effective measures for the control of tomato diseases. Lifetime of the resistant varieties which are usually recommended for production is often limited by the emergence of new races of different pathogens exceeding the resistance of genes of cultivated varieties (Amini, Sidovich, 2010).

Chemical control of diseases is usually effective, but it has some unintended consequences as the reduction of amount of many beneficial organisms, as well as toxic impacts on human health and environment (Manczinger, Antal, Kredics, 2002; Gavrilescu, Chisti, 2005).

From the point of economic view and ecological advantageous the use of varieties with long-term resistance for the control of tomato diseases produced by fungi is preferable.

The aim of our study was to test the level of resistance of new tomato varieties and lines to some fungal diseases (*Fusarium* spp., *A. alternata*) for the selection of forms with enhanced resistance.

MATERIALS AND METHODS

Six varieties and twelve perspective lines of tomato, created in the Institute of Genetics, Physiology and Plant Protection (IGPPP) and manifested the complex of valuable characters, were used as a material for research.

Seeds of tomato were treated with FC of the fungi for 18 hours. Seeds, which were contained in distilled water, were used as a control. The cultivation of seedlings was carried out in Petri dishes on the filter paper, moistened with distilled water, at room temperature 22-24°C for 6 days. The important characters of growth and development of tomato at the early stages of ontogeny –

germination, length of radicle and length of stemlet – were used as index-test of the reaction of plants.

In order to elucidate the features of the action of fungi *F. oxysporum, F. solani, F. redolens* and *A. alternata* on the ability of germination and growth of radicle and stemlet of tomato the analysis Box & Whisker Plot was made.

Cluster analyses were performed by creating dendrograms on the base of agglomerative-iterative algorithm using Ward method and the method *k*-means (Savary, 2010).

Bifactorial analysis of the variance ANOVA was applied to appreciate the role of the genotype, the species of fungus, and their interaction as a source of variation of quantitative traits.

RESULTS AND DISCUSSIONS

Analysis of the reaction of varieties and perspective lines of tomato to seeds treatment with the FC of fungi *F. oxysporum, F. solani, F. redolens* and *A. alternata* isolated from roots and leaves with symptoms of disease allowed to find that the reaction of plants to all four FC was different, specific for genotype, analyzed character, and species of fungus. Responses to infection were classified to the categories: inhibition, stimulation, lack of reaction (Table 1).

Table 1. The influence of culture filtrates of *Fusarium* spp. and *Alternaria alternata* on some characters of tomato growth and development

No.	Variant	Germi-nation, %	% relative to control	Length of radicle, mm	% relative to control	Length of stemlet, mm	% relative to control
				'Tomis'			
1.	H₂O (control)	100.0	100.0	37.7±1.8	100.0	20.8±1.0	100.0
2.	FC *F. oxysporum*	88.3	88.3	32.6±2.4	86.5	15.6±1.2*	75.0
3.	FC *F. solani*	48.3	48.3	27.9±3.9*	74.0	26.8±3.1	128.8
4.	FC *F. redolens*	81.7	81.7	29.6±3.1*	78.5	19.1±2.5	91.8
5.	FC *A. alternata*	78.3	78.3	16.2±1.6*	43.0	12.2±1.2*	58.6
				'Exclusiv'			
1.	H₂O (control)	98.3	100.0	38.3±2.3	100.0	25.4±1.1	100.0
2.	FC *F. oxysporum*	91.5	93.1	15.5±1.3*	40.5	10.2±0.8*	40.2
3.	FC *F. solani*	71.2	72.4	13.6±1.6*	35.5	14.3±2.0*	56.3
4.	FC *F. redolens*	89.8	91.3	25.6±3.1*	66.8	15.8±1.9*	62.2
5.	FC *A. alternata*	100	101.7	31.3±2.6	81.7	18.6±1.5*	73.2
				'Mihaela'			
1.	H₂O (control)	100.0	100.0	37.4±2.0	100.0	19.5±1.2	100.0
2.	FC *F. oxysporum*	90.0	90.0	35.8±3.1	95.7	18.5±1.6	94.9
3.	FC *F. solani*	55.0	55.0	33.3±4.2	89.0	23.3±1.2*	119.5
4.	FC *F. redolens*	88.3	88.3	44.2±3.4	118.2	23.6±1.6*	121.0
5.	FC *A. alternata*	90.0	90.0	31.8±2.3	85.0	16.7±1.3	85.6

No.	Variant	Germi-nation, %	% relative to control	Length of radicle, mm	% relative to control	Length of stemlet, mm	% relative to control
				'Mary Gratefully'			
1.	H_2O (control)	91.7	100.0	43.5±2.7	100.0	26.8±1.3	100.0
2.	FC *F. oxysporum*	92.7	101.1	46.6±92.7	107.2	23.3±1.4	86.9
3.	FC *F. solani*	78.2	85.3	42.0±4.2	96.5	23.3±1.8	86.9
4.	FC *F. redolens*	70.9	77.3	24.0±2.9*	55.2	9.5±1.6*	35.3
5.	FC *A. alternata*	81.8	89.2	11.0±0.8*	25.3	8.6±0.8*	32.1
				'Desteptarea'			
1.	H_2O (control)	85	100.0	40.9±2.9	100.0	27.1±1.9	100.0
2.	FC *F. oxysporum*	76.5	90.0	13.6±1.5*	33.2	12.7±1.7*	46.9
3.	FC *F. solani*	80.4	94.6	21.6±3.0*	52.8	23.8±2.3	87.8
4.	FC *F. redolens*	80.4	94.6	56.3±3.3*	137.6	28.2±1.8	104.1
5.	FC *A. alternata*	100.0	117.6	53.6±3.9*	131.0	26.6±1.7	98.1
				'Milenium'			
1.	H_2O (control)	86.7	100.0	36.9±2.8	100.0	22.9±1.4	100.0
2.	FC *F. oxysporum*	98.1	113.1	13.9±1.0*	37.7	8.9±0.8*	38.9
3.	FC *F. solani*	73.1	84.3	14.0±2.1*	37.9	13.8±3.7*	60.3
4.	FC *F. redolens*	84.6	97.6	27.3±2.7*	74.0	17.2±1.6*	75.1
5.	FC *A. alternata*	100.0	115.3	26.1±2.5*	70.7	15.7±1.5*	68.6

For example, estimating the capacity of seeds germination of approved varieties after treating them with FC of mentioned above fungi we found that varieties 'Exclusiv', 'Desteptarea', 'Milenium' manifested the highest resistance. This index has diminished under the action of these pathogens to 6.9-27.6%; 5.4-10.0%; 2.4-15.7%, respectively, but sometimes significant stimulation was recorded.

Thus, the variety 'Milenium' under the influence of FC *F. oxysporum* demonstrated increasing 13.1%, 'Desteptarea' and 'Milenium' under the influence of FC *A. alternata* – 17.6 and 15.3%, respectively.

The highest complex resistance was recorded at the 'Mihaela' variety by the evaluation of the reaction of radicles and stemlets to the action of FC of four fungi used in research, the phenomenon was manifested as insignificant diminishing or stimulation: -15.0 ... + 18.2% and -14, 4 ... + 21.0%, respectively, for radicles and stemlets.

In order to elucidate the peculiarities of action of fungi *F. oxysporum*, *F. solani*, *F. redolens* and *A. alternata* on the capacity of germination, growth of radicle and stemlet in tomato the analysis Box & Whisker Plot was proceeded, 6 varieties were used as cases (Figure 1).

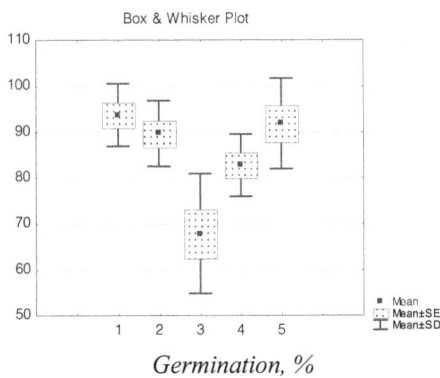

Germination, %

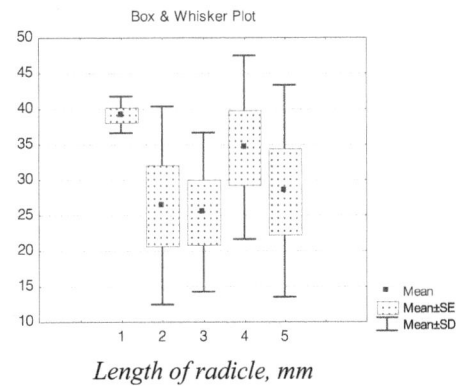

Length of radicle, mm

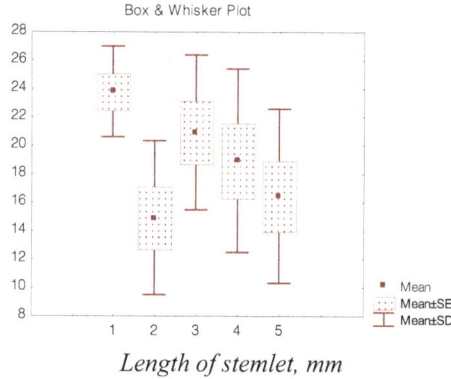

Length of stemlet, mm

Figure 1. The influence of culture filtrates of causative agents of root rot in tomato to the growth and development of
seedlings at the early stages:
1 – control (H_2O), 2 – FC *F. oxysporum*, 3 – FC *F. solani*, 4 – FC *F. redolens*, 5 – FC *A. alternata*

The data obtained showed that in the case of germination the strongest repression (with statistical support, $p \leq 0.05$) was recorded under the influence of fungus *F. solani* (-27.7%) and *F. redolens* (-11.7%), in the case of radicle growth – *F. solani* (-35.1%), and in the case of stemlet growth – *F. oxysporum* (37.4%) and *A. alternata* (-31.0%). The results demonstrated the specificity of action of these pathogenic fungi on growth and development of tomato seedlings at the early stages of ontogeny. From such reason we can conclude that the complex test of perspective tomato forms is necessary in the process of creation of varieties which are capable to develop normal growth organs in the presence of mentioned fungi in the soil.

By cluster analysis some similarities and differences between studied varieties were found when the reaction of germination and growth of seedlings to the influence of fungi *Fusarium* spp. and *A. alternata* was registered (Figure 2).

Thus, similarity of varieties 'Tomis' and 'Mihaela' was shown by analysis of germination, they have diminished indices; 'Exclusiv' and 'Mary Gratefully' – intermediate level; 'Desteptarea' and 'Milenium' – the highest germination and sometimes attested stimulation.

Analyzing growth of radicle and stemlet we found that following variety 'Tomis', 'Mihaela' and 'Mary Gratefully' has demonstrated similarity of these indexes as high. This tells about their fitness for production, provides additional conditions for sowing seeds or seedlings to avoid the danger of reducing their germination capacity in the case of strong soil infestation with causal agents of root rot.

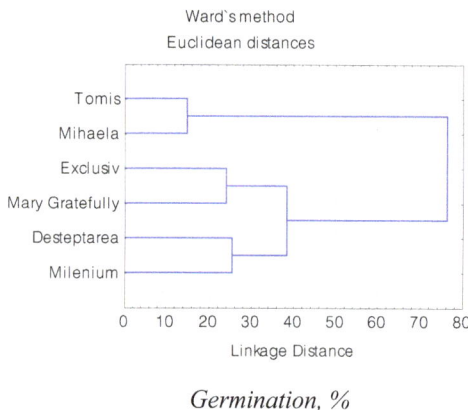

Germination, %

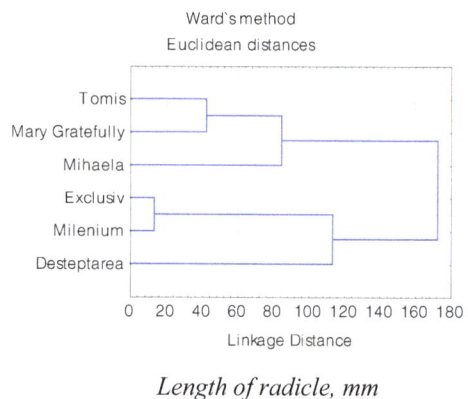

Length of radicle, mm

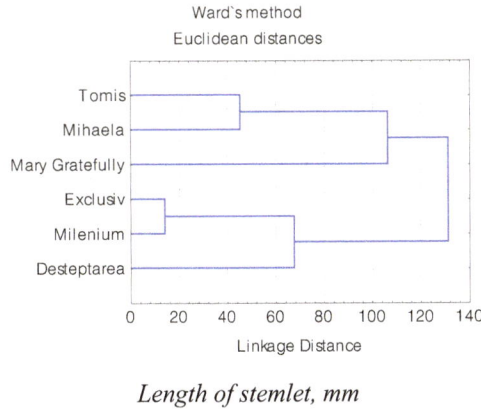

Ward`s method
Euclidean distances

Length of stemlet, mm

Figure 2. The dendrogram distribution of approved tomato varieties on the basis of similarity of responses to pathogens *Fusarium* spp. and *Alternaria alternata*

By bifactorial analysis of variance it was found that the greatest contribution in the variation of length of radicle embryonic and stemlet had a species of fungus, its contribution was 41.07 and 58.58%, respectively (Table 2).

Table 2. Factorial analysis of the *tomato genotype x fungal pathogen* relationships in tomato

Source of variation	Degree of freedom	Mean sum of squares	Contribution in the source of variation, %
Length of radicle			
Tomato genotype	5	8100*	32.49
Species of fungus	4	10241*	41.07
Tomato genotype x species of fungus	20	6242*	25.03
Random effects	1424	351	1.41
Length of stemlet			
Tomato genotype	5	1210.2*	22.00
Species of fungus	4	3222.3*	58.58
Tomato genotype x species of fungus	20	975.1*	17.73
Random effects	1103	93.5	1.70

*-p≤0,05

It was established that the genotype played an important role too; its factorial pondering consisted of 32.49% and 22.00% for the length of radicle and stemlet, respectively; the interaction *tomato genotype x species of fungus* was recorded as 25.03% and 17.73%, respectively, for both characters. The significant pondering of the contribution of fungus species (*Fusarium* spp., *A. alternata*) and interaction *tomato genotype x species of fungus* in the variation of growth plant organs reveals the necessity of constant monitoring of the composition of species of pathogen agents and their virulence.

The 12 lines obtained by intra-species crosses were studied in order to identify tomato lines with increased resistance to pathogens *Fusarium* spp. on the base of reaction to filtrates of fungi cultures, varieties 'Elvira' and 'Trapeza' were used as the control.

In the control variants the **germination** of seeds ranged within 88.3 - 95.0% in most lines excepting two lines L 202 and L 309: 61.7; 65.0% (Table 3).

The results demonstrated that FC of studied fungi influenced different on seeds germination.

For example, inhibition was -5.7 ... -24.2% under the influence of *F. oxysporum* FC. Strong repression was found for 'Trapeza' (-17.0%), L 207 (-22.8%), L 315 (-24.2%), lack of reaction – f or L 202, L 204, L 204, L 206, stimulation: L 208 (+ 9.8%), L 313 (+ 5.3%), L 314 (+ 8.0%), insignificant cases: 'Elvira', L 203, L 308 (-5.7 ... - 9.2%).

In the case of FC *F. solani* inhibition was more pronounced and varied in 1.9 ... 30.5%. Significant inhibition was registered in 'Trapeza' (-30.1%), L 203 (-30.5%), L 308 (23.7%), L 314 (-24.0%),   L 315 (-15.5 %), L 207 (-14.0%), lack of reaction   in the line L 309, stimulation – L 204 (+ 5.3%), insignificant reaction: L 202 (-2.8%), L 204 (5.3%), L 204 (+ 7.3%), L 206 (-3.3%), L 208 (7.9%), 'Elvira' (-1.9%), L 313 (7, 1%).

FC *F. redolens* produced inhibition at 12 forms from 14, it varied within the   ranges -1.9 ... -27.3%, two forms – L 202, L 309 – had no response to the treatment. It should be mentioned that repression was up to 10% in 8 genotypes.

Table 3. The influence of culture filtrates of root rot causative agents to the seeds germination

Line	Origin	Control H_2O	*F. oxysporum*	% relative to control	*F. solani*	% relative to control	*F. redolens*	% relative to control
	Trapeza, control	88.3	73.3	-17.0	61.7	-30.1	75.0	-15.1
202	F_7 Maestro x Irişca	61.7	61.7	0	60.0	-2.8	61.7	0
203	F_6 (Maestro x Irişca) x Maestro	98.3	91.7	-6.7	68.3	-30.5	76.7	-22.0
204	F_6 (Maestro x Irişca) x Irişca	93.3	93.3	0	98.3	+5.3	86.7	-7.1
204a	F_6 (Maestro x Irişca) x Irişca	91.7	91.7	0	85.0	-7.3	85.0	-7.3
206	F_6 (Maestro x D.M.M.) x D.M.M.	100.0	100.0	0	96.7	-3.3	90.0	-10.0
207	F_7 'Mihaela' x Irişca	95.0	73.3	-22.8	81.7	-14.0	83.3	-12.3
208	F_7 'Mihaela' x D.M.M.	85.0	93.3	+9.8	78.3	-7.9	83.3	-2.0
	Elvira, martor	88.3	83.3	-5.7	86.6	-1.9	83.3	-5.7
308	F_{13} Nistru x L 325	91.7	83.3	-9.2	70.0	-23.7	66.7	-27.3
309	F_{11} Nota x Kecskemeti	65.0	55.0	-15.1	65.0	0	65.0	0
313	F_{13} Novicioc x Iuliana	95.0	100.0	+5.3	88.3	-7.1	88.3	-7.1
314	F_{12} Uspeh x L 325	83,3	90.0	+8.0	63.3	-24.0	81.7	-1.9
315	F_{10} Nistru x Saladette	96.7	73.3	-24.2	81.7	-15.5	88.3	-8.7

Radicle embryonic. It was found that genotypes demonstrated very different susceptibility to FC, mean values with the relation to the control varied within the limits -0.4 ...-75.5% for *F. oxysporum* isolate, - 9.6 ... -76.4% – *F . solani*, and -12.4 ... -42.5% – *F. redolens* (Table 4). It should be mentioned that *F. oxysporum* in 9 cases from 14 had stimulating influence (+3.4 ... + 169.0%), L207 manifested a strong sensitivity (-75.0%). Evaluated lines were most strongly influenced by *F. solani*.

Eleven from 14 genotypes exhibited a suppression which was within the limits - 9.6....-76.4%. Stimulation was found in 'Trapeza' (+ 8.4%), L 206 (+ 19.0%) and L 315 (+ 4.5%). In the variant with FC *F. redolens* 7 from 14 genotypes showed inhibition, but 6 lines demonstrated the stimulation of radicle growth. So the lines were strongly different in this analyzed character that reveals the opportunity to identify resistant genotype.

Table 4. The influence of culture filtrates of root rot causative agents to the embryonic radicle length

Line	Origin	Control H_2O	F. oxysporum	% relative to control	F. solani	% relative to control	F. redolens	% relative to control
	Trapeza	40.6±1.88	42.0±1.96	+3.4	44.0±3.58	+8.4	41.0±2.32	+1.0
202	F_7 Maestro x Irişca	27.8±1.91	27.7±2.03	-0.4	9.5±1.18	-65.8	38.2±1.92	+37.4
203	F_5 (Maestro x Irişca) x Maestro	30.3±1.39	35.9±2.00	+18.5	17.3±3.07	-42.9	18.6±1.52	-38.6
204	F_5 (Maestro x Irişca) x Irişca	27.7±1.26	36.8±2.31	+32.8	22.4±2.39	-18.2	32.0±1.97	+15.5
204a	F_5 (Maestro x Irişca) x Irişca	13.6±1.03	36.7±2.38	+169.8	12.3±1.10	-9.6	28.7±1.75	+111.0
206	F_5 (Maestro x D.M.M.) x D.M.M.	40.7±1.88	52.5±1.77	+29.6	48.2±2.15	+19.0	42.6±2.02	+5.2
207	F_7 'Mihaela' x Irişca	37.2±2.05	9.1±0.75	-75.5	15.9±2.45	-57.0	40.82±2.42	+9.7
208	F_7 'Mihaela' x D.M.M.	48.4±2.49	47.2±2.12	-2.5	30.0±3.81	-38.0	30.5±2.14	-37.0
	Elvira	36.0±2.48	47.6±2.42	+32.2	16.5±2.10	-54.2	20.7±2.01	-42.5
308	F_{13} Nistru x L 325	45.1±2.54	54.0±2.14	+19.7	11.8±1.91	-73.8	37.7±2.32	-16.4
309	F_{11} Nota x Kecskemeti	23.8±4.56	23.0±2.24	-3.4	12.9±1.94	-45.8	24.4±3.35	+2.5
313	F_{13} Novicioc x Iuliana	45.9±2.00	48.1±1.95	+4.8	30.7±2.82	-33.1	40.2±2.07	-12.4
314	F_{12} Uspeh x L-325	38.1±1.75	43.1±2.64	+13.1	9.0±0.83	-76.4	22.5±2.04	-41.0
315	F_{10} Nistru x Saladette	35.7±1.71	30.8±2.09	-13.7	37.3±1.83	+4.5	29.7±1.33	-16.8

Stemlet. Its repression varied within the limits -0.5...-74.3% for *F. oxysporum*, -16.7... -74.5% – *F. solani*, -5.2...-56.6% – *F. redolens* (Table 5). Strong sensitivity to *F. oxysporum* was recorded in lines L 207 (-74.3%), L 208 (33.3%), L 202 (-19.2%), L 315 (-26.8%), stimulation was registered in 5 lines: L 203 (+ 5.8%), L 204 (+ 11.1%), L 204a (+ 58.5%), L 308 (+ 23.8%), L 314 (+ 1.6%) (Table 5). Filtrate of culture of *F. solani* in 10 cases from 14 inhibited stemlets growth.

High sensitivity was found in lines: L 202 (-74.5%), L 206 (26.1), L 308 (-64.9%), L 309 (-50.0%), L 314 (- 42.8), stimulation in the variety 'Trapeza' (+ 1.4%); L 203 (+ 5.8%), L 204 (+ 47.7%), L 204a (+ 3.8%).
Thus, the lines L 203, L 208 and variety 'Elvira' were the most sensitive to *F. redolens,* repression consisted of 46.8; 56.6 and 40.2%, respectively. Strong stimulation was found in L 309 (52.5%).

Table 5. The influence of culture filtrates of root rot causative agents to the tomato stemlet length

Line	Origin	Control H_2O	F. oxysporum	% relative to control	F. solani	% relative to control	F. redolens	% relative to control
	Trapeza	21.2±1.03	20.2±1.27	-4.7	21.8±1.57	+1.4	25.0±1.72	+17.9
202	F_7 Maestro x Irişca	18.8±1.30	15.2±1.33	-19.2	4.8±1.21	-74.5	20.6±1.32	+9.6
203	F_6 (Maestro x Irişca) x Maestro	18.8±1.06	19.9±1.31	+5.8	19.9±3.51	+5.8	10.0±0.90	-46.8
204	F_6 (Maestro x Irişca) x Irişca	15.3±0.70	17.0±0.95	+11.1	22.6±1.64	+47.7	17.2±1.43	+12.4
204a	F_6 (Maestro x Irişca) x Irişca	13.0±1.17	20.6±1.52	+58.5	13.5±2.94	+3.8	14.0±0.85	+7.7
206	F_6 (Maestro x D.M.M.) x D.M.M.	31.4±1.07	26.6±0.85	-15.5	23.2±1.05	-26.1	23.6±1.39	-24.9

Line	Origin	Control H_2O	F. oxyspo-rum	% relative to control	F. solani	% relative to control	F. redolens	% relative to control
207	F_7 'Mihaela' x Irişca	27.2±1.26	7.3±1.31	-74.3	21.8±3.24	-19.9	25.8±1.56	-5.2
208	F_7('Mihaela' x D.M.M.)	28.8±1.88	19.2±1.01	-33.3	23.3±1.57	-19.1	12.5±1.31	-56.6
	Elvira	20.4±1.38	20.5±1.25	-0.5	17.0±1.83	-16.7	12.2±1.26	-40.2
308	F_{13} Nistru x L 325	26.5±1.25	32.8±1.41	+23.8	9.2±2.78	-64.9	18.8±1.58	-29.1
309	F_{11} Nota x Kecskemeti	11.8±3.16	11.7±1.37	-0.9	5.9±1.45	-50.0	18.0±2.74	+52.5
313	F_{13} Novicioc x Iuliana	28.4±1.04	27.5±1.53	-3.2	22.9±1.61	-19.4	21.2±1.47	-25.4
314	F_{12} Uspeh x L 325	18.7±1.13	19.0±1.34	+1.6	10.7±3.12	-42.8	12.1±1.05	-35.3
315	F_{10} Nistru x Saladette	27.2±0.94	19.9±1.45	-26.8	21.0±0.97	-22.8	20.5±1.25	-24.6

Cluster analysis (*Ward method*) allowed finding similar particularities in created varieties and lines with respect to studied characters: germination, length of radicle and stemlet (Figure 3).

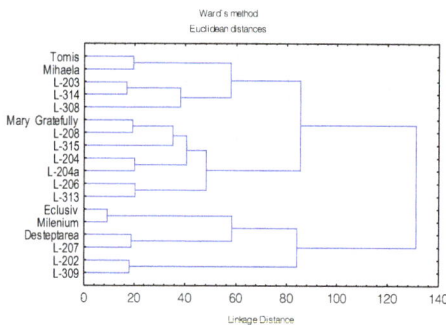

Figure 3. The degree of similarity of tomato varieties and lines on the base of the reaction to pathogens *Fusarium* spp.
(*Cases*: germination, length of radicle, length of stemlet)

base of different reaction to FC *Fusarium* spp. (Figure 4).

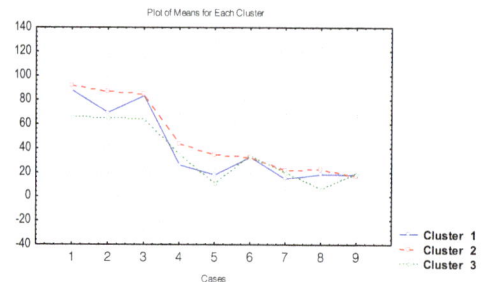

Figure 4. Cluster analysis (*k*-means) of distribution of genotypes on the base of response to culture filtrates of *Fusarium* spp.
Horizontal: germination (% to the control) under the FC influence : 1 – *F. oxysporum, 2 – F. solani, 3 – F. redolens; length of radicle*: 4 – *F. oxysporum, 5 – F. solani, 6 – F. redolens; length of stemlet*: 7 – *F. oxysporum, 8 – F. solani, 9 – F. redolens.*
Vertical: 1, 2, 3 – clusters of genotypes

Clusters of genotypes with similar response to pathogens *Fusarium* spp. were identified. It was found that the lines L 203, L 208, L 314 were high similar with the varieties 'Tomis' and 'Milenium', and L 202, L 207, L 209 – with varieties 'Exclusiv', 'Milenium' and 'Desteptarea'.
Cluster analysis of *k-means* (*method centroid*) demonstrated that groups of varieties and lines were separated into three clusters on the

As members of the *cluster 1* were: 'Tomis', 'Exclusiv', 'Mihaela', 'Desteptarea', 'Milenium', L 203, L 204a, L 207, L 314; *cluster 2:* 'Mary Gratefully', L 204, L 206, L 208, L 313, L 315; *cluster 3*: L 202, L 308, L 309. The varieties of clusters 1 and 2 showed higher resistance compared with the cluster.

CONCLUSIONS

It has been found that the pathogens *Fusarium* spp. and *A. alternata* significantly influenced on the early ontogeny of tomato genotypes by the suppression of seeds germination, growth of radicle and stemlet (sometimes - by their stimulation). Cluster analysis using *k-means* method allowed to find that fungal species *F. solani* and *F. redolens* showed a higher discriminating capacity on varieties / lines of tomato in comparison with *F. oxysporum*. This reveals more pronounced specificity of interaction with these pathogens.

The varieties 'Exclusiv', 'Desteptarea', 'Milenium' manifested the highest resistance in seeds germination capacity, as well as the variety 'Mihaela' demonstrated the highest complex resistance in respect with the reaction of radicle and stemlet to all 4 culture filtrates.

Using factorial analysis it was found that the species of fungus (*Fusarium* spp., *A. alternata*) included the greatest contribution to the source of variation of radicles and stemlets length, its contribution consisted of 41.07 and 58.58%, respectively.

The significant pondering of the influence of fungus species (*Fusarium* spp., *A. alternata*) and the interaction tomato genotype x species of fungus on the source of variation of plant growth organs reveals the necessity of constant monitoring of the composition and virulence of fungal species causing root rot in tomato.

REFERENCES

Amini J., Sidovich D.F., 2010. The Effects of Fungicides on Fusarium oxysporum f. sp. lycopersici Associated with Fusarium Wilt of Tomato. In: J. of Plant Protect. Research, vol. 50, nr. 2, 172-178.

Foolad M.R., 2007. Genome mapping and molecular breeding of tomato. In: International J. of Plant Genomics, 52.

Gavrilescu M., Chisti Y., 2005. Biotechnology – a sustainable alternative for chemical industry. In: Biotechnol. Adv., vol. 23, 471-499.

Manczinger L., Antal Z., Kredics L., 2002. Ecophysiology and breeding of mycoparasitic Trichoderma strains. In: Acta Microbiologica et Immunologica Hungarica, vol. 49, 1-14.

Polixenova V.D., 2009. Induced resistance of plants to pathogens and abiotic stress factors (in the example of tomato). Annals of BSU, Ser. 2, N 1, 48-60.

Savary S. et al., 2010. Use of Categorical Information and Correspondence Analysis in Plant Disease Epidemiology. In: Adv. in Bot. Research, vol. 54, 190-198.

Scott J.W., Gardner R.G., 2007. Breeding for Resistance to Fungal Pathogen. In: Genetic Improvement of Solanaceous Crops. Vol. 2. Tomato. Ed. Razdan M.K., Mattoo A.K., USA, Sci. Publishers, 421-456.

Yamagishi D. et al., 2006. Pathological evaluation of host-specific AAL-toxins and fumonisin mycotoxins produced by *Alternaria* and *Fusarium* species. In: J. of General Plant Pathol., Vol. 72, Issue 5, 323-327.

10

STUDY ON THE INFLUENCE OF SUBSTRATE CULTURE ON THE PRODUCTION OF CUCUMBERS IN UNCONVENTIONAL SYSTEM

Petre Sorin[1], Maria PELE[1], Elena Maria DRĂGHICI[2]

[1]University of Agronomic Sciences and Veterinary Medicine of Bucharest – Faculty of Biotechnologies, 59 Mărăşti Blvd, District 1, 011464, Bucharest, Romania
Email: mpele50@yahoo.com

[2]University of Agronomic Sciences and Veterinary Medicine of Bucharest – Faculty of Horticulture, 59 Mărăşti Blvd, District 1, 011464, Bucharest, Romania
Email: elena.draghici@horticultura-bucuresti.ro

Corresponding author email: elena.draghici@horticultura-bucuresti.ro

Abstract

Cucumbers are one of the basic crops in greenhouse being grown on large areas in both the cycle and cycle II of culture. It is a very demanding species to the conditions of culture, but if you apply an adequate technology can bring obtained yields with important benefits. In greenhouse is practiced both the culture on soil and soilless culture. The advantages of cultivation on soilless consist of an effective monitoring of medium of culture, in particular irrigation and nutrition. Culture substrates are selected such that they do not interact with the nutritional solution. Perlite is a substrate inexpensive, reusable and ensuring earliness and production increase. The present study was conducted in greenhouses at Hortinvest Research Centre of University of Veterinary Medicine from Bucharest. Culture of cucumber was established in the first cycle and used Pyralis hybrid, specific for greenhouse crop. We used three experimental variants of mattresses filled with grain of 2, 4 and 5 mm diameter. The best solution to a grain size of perlite was the variant with 4 mm diameter, ensuring the most satisfying results for early and total production. The aim of the study was to identify the best solution culture substrate games and recommend the use of crops without soil growers.

Key words: perlite substrate, cucumbers, size grain.

INTRODUCTION

In Romania, cucumbers culture occupies an important place. The cucumbers are cultivated in different growing systems, as: greenhouse, solar, polytunnels or open field, so the production is covered market throughout the year. Growing cucumbers on nutrient substrate is practiced only on farms that have appropriate technology as this type involves careful coordination of all environmental factors, in particular the fertigation.

FAO Yearbook 2012 states that in 2004-2011, the production and the cultivated areas with cucumbers had had on all continents a significant increment. Increased production was based on improving production efficiency as a result of technical progress in this area, diversification assortment of varieties grown, expanding culture of hybrids with high yield potential, reduce losses caused by pests and diseases through integrated control of their sector developer of greenhouses and solariums by increasing the surface and generalization of modern technologies, concentration of production in favourable areas.

In recent years, many studies have been made on soilless cucumber production in greenhouses in Turkey (Özgür, 1991; Canatar, 1997; Saracoglu, 1997; Öztan, 2002; Kaptan, 2006; Gül et al., 2006; Gül et al., 2007. cited by Engindeniz and Gül, 2009). Though, there is still need for study, especially on economics of soilless cucumber production at farmers' level.

Therefore, the researchers are permanently constrained to finding new modern growing technology, perfumed that to assure a high production.

The most frequently unconventional system is the system of cucumber growing on substrate of Grodan (Petre et al., 2014).

In view of the above, it is necessary to develop technologies that are not expensive, can be made with cheap materials and handy, but at the same time ensuring high productivity both quantitatively and qualitatively.

The culture of perlite substrate has two major advantages: it is very accessible from economically within the global trend as organic.

Results made of Peyvast et al., 2010, showed that substrates had a significant effect on the plant growth, total fruit yield, marketable fruits, fruit weight and number of fruits per m^2.

In the global horticultural production, vegetable crops "without soil" had begun already gain a leading position. These unconventional systems of culture are great interest both for researchers and for those who practice in order to achieve products for human consumption.

In Romania, expansion of these systems raises serious technical and economic issues, so it is necessary to establish culture technologies applicable, using local materials and equipment imported or to be accessible to a larger number of users.

Extending this systems create some problems referring to polluting because the Grodan is a substrate that is difficult to recycled.

Purpose of research in this study was to identify the best composition based on perlite substrate and recommend it to obtain early and high yields, quality and price of low cost. Expanded perlite is a substrate of culture that completely replaces soil.

MATERIALS AND METHODS

The study was conducted within the greenhouses of Hortinvest Research Center – University of Agronomic Sciences and Veterinary Medicine Bucharest, between February and June 2014. The biological material used was the cucumber hybrid Pyralis.

The experiments consisted in cultivation on the Perlite substrates presented in Table 1 and monitoring various growth factors.

The culture was established in greenhouse heated to 10 February 2014. The planting seedlings had 32 days.

Table 1. Experimental variants

Variants	Substrate types	Growing
V1	Perlite 2mm	Growing on mattresses
V2	Perlite 4mm	Growing on mattresses
V3	Perlite 5mm	Growing on mattresses

Of each variant we use twenty four mattresses of 1 m long for each where we had planted each three plants. For each plant has been assured 10 l perlite substrate. Mattresses had contained 30 l of substrate. Plant density was 18,500 plants per ha.

Hydroponics mattresses were made of biodegradable polyethylene, triple laminated, composed of two layers, colored black inside and white outside. Mattresses have a length of 1 m and a width of 20 cm

The fertilizing recipe was modified according to phenophase.

In the first phenophase, immediate period after planting, for each plant were provided an amount of 50 ml of solution per fertigation - for 2 weeks.

Daily it has been administrated a number of six watering.

The amount of solution per plant as the plants increased in height was increased, so had administrated between 150 and 200 ml depending on temperature and light.

During the growing season were conducted observations and determinations so:
- Plant growth in height;
- Early production;
- The quantity of fruit harvested per plant;
- The average fruit per harvest;
- The total production;

Fruit production was determined by weighing. Each assay was performed at least 3 times.

For each determination was made statistical analysis

RESULTS AND DISCUSSIONS

During the period of culture the greenhouse temperatures were recorded and the averages are shown in Figure 1.

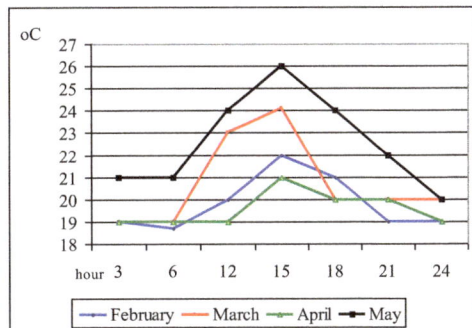

Fig. 1. Average temperature recorded in greenhouse

The plants had grown in height between 163.8 cm and 171.2 cm V1 to V3 (table 2).

Table 2. Dynamic growing in height of cucumber plants

Variant	UM	Number days after planting			
		10	20	30	40
V1	cm	26.8	127.5	147.5	163.8
V2	cm	24.5	117.1	140.7	165,0
V3	cm	24.3	122.8	143.2	171.2

From the statistical point of view, we could notice insignificant differences in plant height growth after ten days of planting (table 3).

In March, the largest quantity was harvested from V2 (1804 kg / plant). The lowest production was obtained at V3 (1,155kg / plant).

Table 3. Summary of results of differences in height after 10 days of planting

Variant	Hight	Difefference	Significance
		(cm)	(cm)	(%)
V(0) Average	25.20	0.00	100.00	Mt
V(1)	26.80	1.60	106.35	N
V(2)	24.50	-0.70	97.22	N
V(3)	24.30	-0.90	96.43	N
DL5% =	1.750	DL5% in % =	6.9444
DL1% =	3.800	DL1% in % =	15.0794
DL01% =	12.890	DL01% in %=	51.1508

Also, we could notice, insignificant differences regarding plant height growth of point of view statistically, table 4.

Table 4. Summary of results of differences in height after 40 days of planting

Variant	Hight	Difference	Significance
	cm)	(cm)	(%)
V(0) Average	166.67	0.00	100.00	Mt
V(1)	163.80	-2.87	98.28	N
V(2)	165.00	-1.67	99.00	N
V(3)	171.20	4.53	102.72	N
DL5% =	7.010	DL5% in % =	4.2060
DL1% =	15.220	DL1% in % =	9.1320
DL01% =	51.550	DL01% in %=	30.9300

In April, we recorded the highest production of 7.811kg / plant at V2 followed by V1 with 7.166 kg / plant and V3 with a production of 6.431 kg / plant.

In May it were harvested from V2 the quantity of 5.249 kg / plant, and from V1 only 4.914 kg / plant (figure 2).

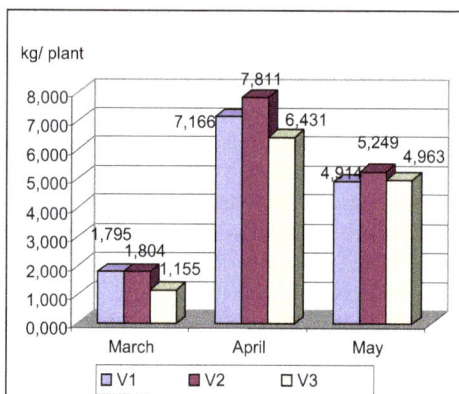

Figure 2. The production obtained per plant

The biggest total production per plant was obtained at V2 (14.864 kg/plant), followed by V1 (13.875 kg/plant). The lowest production I got it from V3 of 12.550 kg / plant (figure 3).

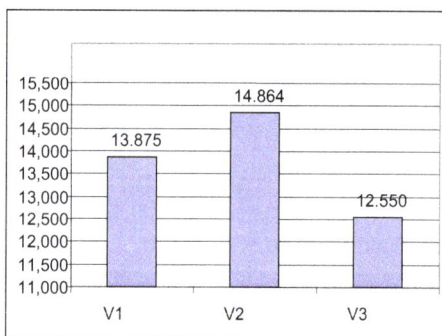

Figure 3. Total yield obtained per plant -kg/plant

CONCLUSIONS

Results show that both in terms of total and per plant production, the best culture substrate is Perlite 3. Reduced particle size of Perlite 1 cause a reduced aeration to roots level and therefore a slow development respectively a lower production. In the case of Perlite 3, the larger size of grains leads to a more rapid drain of fertilizer solutions and therefore a lower consumption of nutrients by plants.

The earliest production was found in the cultures developed on Perlite 3, too.

It has to be emphasized that a problem in greenhouse culture is the temperature variations, because the outside temperature has a real influence on inside temperature even though it is used a good system to keep the temperature at the same value. So, evaluation of temperature variations shows a clear increase in temperature in the greenhouse in May. The explanation is in the fact that the warm weather outside cause an increase in the temperature inside the greenhouse. Therefore it requires a temperature control of greenhouse temperature through the use of appropriate ventilation systems. Of cause, this action will lead to a relatively high consumption of electricity.

Thus, as a general conclusion, the best substrate of those tested for growth of cucumbers in greenhouses is Perlite 3.

ACKNOWLEDGEMENTS

This work was supported by a grant of the Romanian National Authority for Scientific Research, CNDI – UEFISCDI, financed from project number PN-II-PT-PCCA-2011-3.2-1351 - Contract No.68/2012 and the European Social Found, Human Resources Development Operational Programme 2007-2013, project no. POSDRU/159/1.5/S/13276.

REFERENCES

Engindeniz S., Gül A., 2009, Economic analysis of soilless and soil-based greenhouse cucumber production in Turkey, Sci. agric. (Piracicaba,Braz.) vol.66 no.5 Piracicaba Sept./Oct.

Petre S., Pele M, Draghici M.E, 2014, Influence of using perlite and eco fertilizers for hydroponic cucumbers culture, Journal of Biotechnology, Vol. 185, Supplement, September 2014, S81, ISSN 0168-1656.

Peyvast, G.H., Olfati, J.A., Ramezani Kharazi, P. and Noori Roudsari, O., 2010, Effect of substrate on greenhouse cucumber production in soilless culture. Acta Hort. (ISHS) 871: 429-436, http://www.actahort.org/books/871/871_59.htm

11

EVALUATION OF TEMPERATURE DATA USAGE THE METHOD OF DEGREE-HOUR IN GREENHOUSES: PEPPER PLANT CASE

Atilgan ATILGAN[1], Ali YUCEL[2], Cagatay TANRIVERDİ[3], Hasan OZ[1], Ahmet TEZCAN[4]

[1]Suleyman Demirel University, Agriculture Faculty, Agricultural Structure and Irrigation Department, 32260 Isparta, Turkey

[2]Osmaniye Korkut Ata University, Osmaniye Vocational School, 80000 Osmaniye, Turkey

[3]Kahramanmaras Sutcu Imam University, Agriculture Faculty, 46100 Kahramanmaraş, Turkey

[4]Akdeniz University, Agriculture Faculty, Agricultural Structure and Irrigation Department, 07100 Antalya, Turkey

Corresponding author email: ctanriverdi@ksu.edu.tr

Abstract

Greenhouses where the production is carried out and optimum environmental conditions achieved according to many plant species are the agricultural structures. Degree-hour or degree-day methods can be used to get knowledge about energy requirements of any buildings. These methods can provide information about on the amount of energy by using measured and meteorological values and obtaining the heating and cooling values of buildings. The study was chosen in the Kumluca district of Antalya, the place where the most intensive of greenhouse cultivation was made in our country. Autumn cultivation is being done for the pepper plant. This study was conducted between October 2015 and June 2016 in plastic greenhouse. The temperature inside the greenhouse values were recorded during the growing period of the pepper plant. The obtained values were analyzed by Student-t test. The regression coefficients were obtained by calculating the prediction equations between the basic temperature values and the heating and cooling degree-hour values. These coefficients are found between 0.997 and 0.999 for heating degrees-hours and between 0.982 and 0.997 for cooling degrees. As a result, efficiency and quality can be increased through air conditioning improvements in the greenhouses located in mild climate areas such as Antalya, where heating is maintained for frost protection and ventilation difficulties are experienced.

Key words: pepper, greenhouse, degree-hour, temperature.

INTRODUCTION

Greenhouses are the agricultural production facilities where production is planned based on technical and economic factors with a view to ensuring the right air conditioning and production planning required for the type of plant to be cultivated.

Having a history of more than 50 years, greenhouse cultivation is conducted on nearly 66,000 hectares of land throughout Turkey, primarily in the Mediterranean Region as well as the Aegean, Marmara and Black Sea regions (Olgun, 2011; Anonymous, 2016).

Antalya is the city in the Mediterranean Region where greenhouse cultivation is most extensively conducted. Kumluca county in the province of Antalya, however, provides the biggest greenhouse cultivation. When the cultivation of the pepper plant in Kumluca area is compared to the overall greenhouse cultivation in Turkey in 2015, it is observed that while the entire greenhouse cultivation of peppers in Turkey took place in nearly 1000 hectares of land, the area devoted to the production of peppers in Kumluca accounted for 30% of the entire production area (300 hectares), which represented 34% (36,000 tons) of the total national production of 105,000 tons. In the same way, of the entire domestic plastic greenhouse cultivation of peppers performed on a land of 3580 hectares, 755 hectares (21%) of it is performed in Kumluca.

Of the 301,600 tons of peppers produced in the plastic greenhouses in Turkey, 90,600 tons

(30%) are produced in Kumluca county (Anonymous, 2016).

Pepper is a plant that is quite fond of light yet insensitive to the duration of daylight. Although it is capable of germinating at above 8°C, the best germination is achieved at 21°C to 28°C.

While the ideal environmental temperature is between 18-23°C, during day and night, when it is a seedling, care should be taken to ensure that night temperature is not below 12°C. While the daily temperature can be allowed to rise to 25°C on sunny days, ventilation should nevertheless be performed when the tempera-ture hits 30°C. Cultivation is ceased fully at 45°C (Sevgican, 1999).

Planning on heating, cooling and ventilation systems in greenhouses are related to the outside air conditions.

While planning such systems, reliance on long term climate data rather than the climate conditions of several years will help planners adopt a more realistic approach in determining future results (Ileri and Uner, 1998). Degree-day values are one of the most basic measurement units used in the estimation of annual energy requirements of a building located in any place or location (Bayram and Yesilata, 2009).

This study aims to identify the relationship between the indoor temperature and the values of heating and cooling degree-hour values in greenhouse cultivation.

To this end, the indoor temperature values measured inside the greenhouse have been calculated and interpreted based on the heating and cooling durations during the cultivation period of peppers through degree-hour method.

MATERIALS AND METHODS

This study was conducted in Kumluca county, Antalya, where pepper production is extensively performed, in a 4 x 19.5 x 62 m size plastic greenhouse (Figure 1).

It was conducted during the dates between October 2015 and June 2016, a period considered to be the cultivation period of peppers. The temperature was measured by using four TESTO 175 H1 brand temperature and moisture meter sensor.

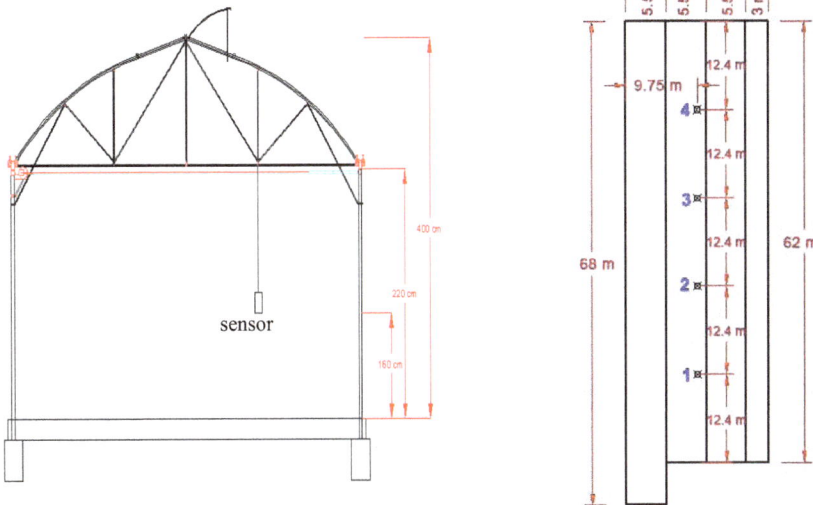

Figure 1. View of the sensors and plastic greenhouse that used in this study

Degree-hour method

Degree-hour and degree-day methods are used in the calculation of heating and cooling loads of structures and air conditioning systems. Degree-hour method yields more precise results than degree-day method (Pusat et. al., 2014).

In the degree-hour method, the energy required for heating or cooling a structure is in proportion with the difference between the air temperature and the balance point temperature. The heating process will be required when the air temperature (T_a) drops below the balance point temperature (Tb). The cooling process will be required when the air temperature (Ta) rises over the balance point temperature (Tb). Heating degree-hour (HDH) and cooling degree-hour (CDH) values can be calculated based on the following equations (Buyukalaca et al., 2001; Bulut et al., 2007; Pusat et al., 2014; Pusat et al., 2015).

$$HDH = (1hour)\sum_{I=1}^{n}\left(T_b - T_a\right)^+ \quad\quad (1)$$

$$CDH = (1hour)\sum_{I=1}^{n}\left(T_a - T_b\right)^+ \quad\quad (2)$$

Here; T_a represents the ambient temperature (°C), T_b the balance point temperature (°C), n the hours of the year, the mark in the equations above suggests that only the positive values will be used.

In the calculation of heating and cooling hours of the pepper plant, the temperature values for autumn cultivation was taken as the basis (Table 1) (Ozalp et al., 2006; Anonymous, 2016).

Table 1. Suggested temperature values for autumn period of pepper plant

Growth period	Suggested temperatures	Date	Period
Greenhouse planting	22-28 °C	The end of September- Beginning of October	7-14 days
Flower formation, pollination, insemination	20-25 °C	2nd of October week - 4th of October week	5-6 weeks
Fruit ripening	16-25 °C	2nd of November week -4th of of December week	6 weeks
Harvest	20-35 °C	2nd of February week- 1st of May week	10 weeks

RESULTS AND DISCUSSIONS

The heating degree-hour graphs are presented in the Figure 2 by taking into account the recommended temperature values for each cultivation period of the pepper plant (planting, flower formation, pollination, insemination, fruit ripening and harvest). The graphs suggest that the heating degree-hour values increase lincarly in accordance with the increasing temperature requirement of the pepper plant in every period. In their study, Bulut et al. (2007) have pointed out that the heating and cooling day values increase linearly depending on the basic temperature values.

The linear increase observed in this study concurs with the findings of Bulut et al. (2007). According to the temperature values measured by the sensors during each cultivation period inside the greenhouse, the calculated numbers of HDHs are found to be fairly close to one another.

It can be argued that there is a homogenous temperature distribution maintained inside the greenhouse. Because heating is maintained in mild climates only for the purpose of protecting the plant against frost and thus heaters are used

(Büyüktas et al., 2016). In the greenhouse where the present study was conducted heating was maintained through heaters and it was stated that this practice was exclusively intended for protecting the plant against frost. Availability of HDH values at every heating requirement throughout each cultivation period of the pepper plant and the fact that they follow a different course than that of the recommended temperature values suggest that the greenhouse was not being heated sufficiently.

Insufficient heating in a greenhouse, on the other hand, leads to a significant decrease in both yield and quality (Sevgican, 1999).

The graphs provided herein present the equations and regression coefficients of different base temperature values recommended throughout the cultivation period (independent variable) and the HDH values (dependent variable).

Upon reviewing such coefficients, it has been established that the regression coefficient (R^2) between the dependent and independent variables vary between 0.996 and 0.999.

This can be explained by the presence of a highly positive relationship.

Greenhouse Planting

Flower formation, pollination, insemination

Fruit ripening

Harvest

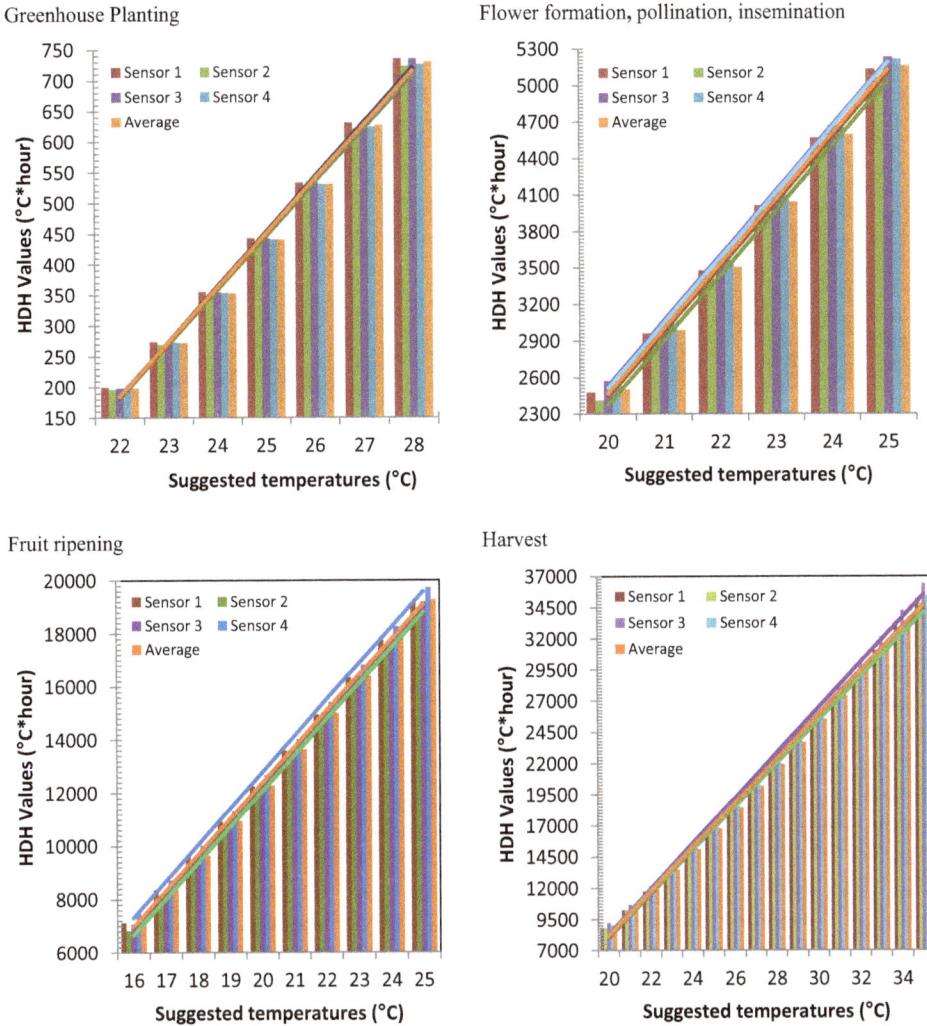

Figure 2. Heating degree-hour values dependent on different temperature values for the pepper plant

The CDH values of the pepper plant have been calculated and presented in the Figure 3.

The CDH values were found to be at a maximum of the values where the heating requirement is reduced for the pepper plant at each cultivation period.

The heating requirement of 22°C during planting, 20°C during florescence, 16°C during fruit development and 20°C during harvesting were determined as the maximum CDH values. The greenhouse, in its current condition, is capable of providing such temperature values.

The fact that the daytime temperature values are higher than that of the night time temperature can be explained by the availability of the CDH values even at the minimum temperature requirements required for the pepper plant.

In their study, researchers (Bayram and Yesilata, 2009; Yucel et al., 2014) state that the numbers of heating degree-days and cooling degree-days are important in terms of determining the capacity and costs of the heating and cooling systems.

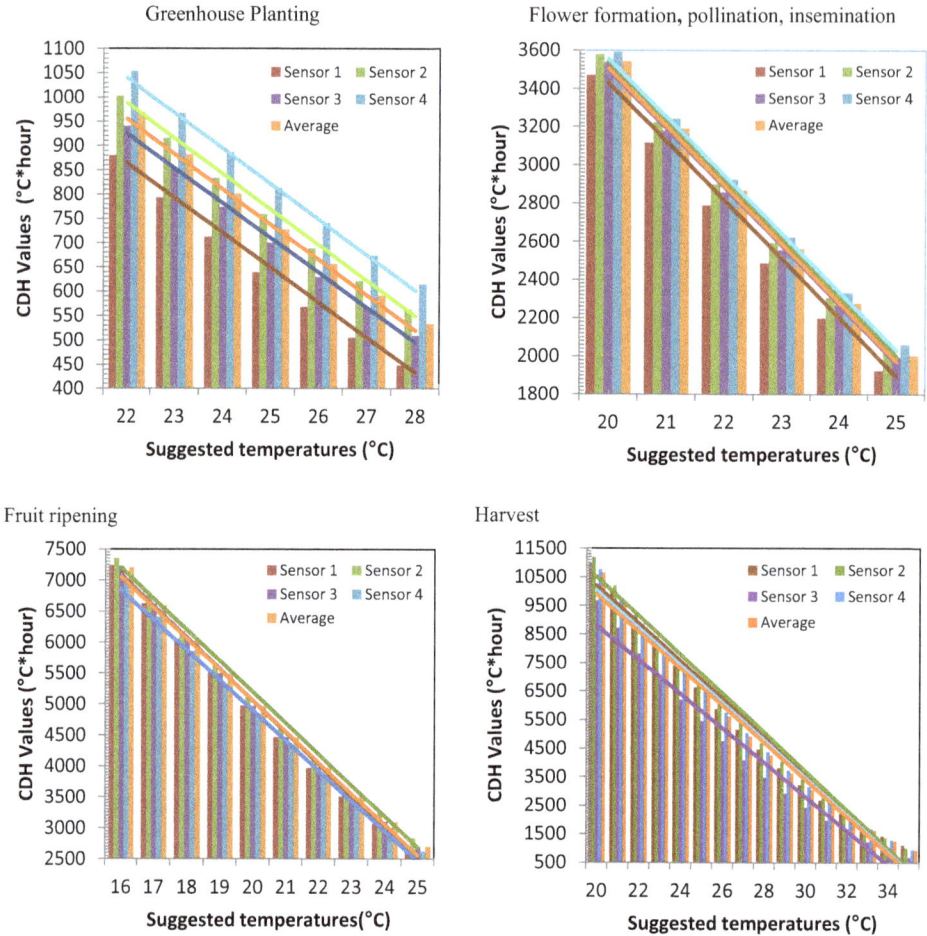

Figure 3. Cooling degree-hour values dependent on different temperature values for the pepper plant

It is reported that the biggest problem encountered in plastic greenhouses is the issue of ventilation. In their study, Monetero et al. (2001) and Fatnassi et al. (2009) report that there are serious efficiency and quality related issues in the greenhouses located around the Mediterranean area, especially at the beginning of the spring and end of the autumn, due to high indoor temperature and insufficient ventilation.

Despite the ventilation provided through both the side walls and the roof, the greenhouse in which the present study was conducted was found to be insufficiently cooled based on the CDH values.

This can help explain the higher cooling degree-hour values even when the heating is least required with the rise in the outside temperature and in line with the cultivation periods of the pepper plant. Again, the equation and regression coefficients of the dependent and independent variables obtained based on the Figure 3 are presented. It is established that such coefficients (R^2) vary between 0.972 and 0.998.

This can also be explained by the presence of a highly positive relationship.

Based on the heating and degree-hour values, we can argue that the plastic greenhouse where this study has been conducted has failed, in its present condition, to provide the temperature values required for the cultivation of peppers at an optimum level. The recommended temperature value of the pepper plant at any cultivation period can be explained by the presence of both heating and cooling values.

CONCLUSIONS

This study has been conducted in a plastic pepper cultivation greenhouse in Kumluca, Antalya where pepper cultivation is extensively performed. Heating and cooling degree-hour values have been determined as per the recommended temperature values for each cultivation period of the pepper plant. Both the heating and cooling degree-hour values were found to have varied based on the temperature requirement of the pepper plant. It was ascertained from the greenhouse growers in the area that the heating was maintained mostly for the purpose of protecting the plant against frost, and it was verified based on the heating and cooling degree-hour values that both heating and cooling was insufficiently maintained for each cultivation period of the pepper plant due to ventilation problems encountered parti-cularly in block plastic greenhouses. In conclusion, yield and quality can be increased through air conditioning improvements in the greenhouses located in mild climate areas such as Antalya, where heating is maintained for frost protection and ventilation difficulties are experienced. Moreover, the fluctuation in the heating and cooling degree-hour values depen-ding on different cultivation periods of the pepper plant may give an idea to the manufacturers beforehand in terms of energy consumption or use.

REFERENCES

Anonymous, 2016. Pepper Growing (*Capsicum annum* L.). http://www.bahcenet.com/biber-yetistiriciligi-capsicum-annum-l.html. (In Turkish)

Bayram M., Yesilata, B., 2009. Integration of number of heating and cooling degree days, IX. National Plumbing Engineering Congress, May 6-9, 2009, Izmir, 425-432. (In Turkish)

Bulut, H., Buyukalaca, O., Yilmaz, T., 2007. Analysis of Heating and Cooling Degree-hour Values for the Mediterranean Region, 2. National Congress of Air conditioning, 111-122.

Buyukalaca, O., Bulut, H.,Yilmaz, T., 2001. Analysis of variable-base heating and cooling degree-days for Turkey, Applied Energy, 69(4):269-283.

Büyüktas, K., Atilgan, A., Tezcan, A., 2016. Agricultural Production Structures, Suleyman Demirel University, Faculty of Agriculture, Publication no: 101, 251. (In Turkish)

Fatnassi, H., Leyronas, C., Boulard, T., Bardin, M., Nicot, P., 2009. Dependence of greenhouse tunnel ventilation on wind direction and crop height. Biosystems Engineering, 103: 338-343.

Ileri, A., Uner, M., 1998. Typical climate data for Turkish cities, IV. National Installation Engineering Congress and Exhibition, 463: 31-42. (In Turkish)

Montero J.I, Hunt, G.R, Kamaruddin, R., Anton, A., Bailey, B.J., 2001. Effect of ventilator configuration of wind-driven ventilation in a crop protection structure for the tropics. Journal of Agricultural Engineering Research 80:99–107.

Olgun, M., 2011. Agricultural Structures, Ankara University publications No:244, 445p. (In Turkish)

Ozalp, R., Çelik, I., Coskun, A., 2006. Covering pepper growing. Sound of agriculture, March 2006, 9:18-21. (In Turkish)

Pusat, S., Ekmekci, I., Dundar, A.C., Ermis, K., Sen, Y., 2014. Typical meteorological year and degree-hour account for Istanbul, 2. National Symposium on Heating-Cooling Education and Exhibition, 23-25 October 2014, 5, Balıkesir. (In Turkish)

Pusat, S., Tunc, N., Ekmekci, I., Yetisken, Y. 2015. Degree-Time calculations for Karabük, ISITES-2015 Valencia-Spain, Academic Platform, 898-905. (In Turkish)

Sevgican, A., 1999. Protected Cultivation, University of Ege, Agriculture Publications Publ, No: 528, 302p. (In Turkish)

TUİK, 2016. Turkish Statistical Institute, www.tuik.gov.tr (In Turkish)

Yucel, A., Atilgan, A., Oz, H., Saltuk, B., 2014. The determination of heating and cooling day values using degree-day method: Tomato plant example. Infrastructure and Ecology of Rural Areas. Nr VI/I, 1049-1061.

12

EFFECT OF THE AGE AND PLANTING AREA OF TOMATO (*SOLANUM LYCOPERSICUM* L.) SEEDLINGS FOR LATE FIELD PRODUCTION ON THE VEGETATIVE BEHAVIOUR OF THE PLANTS DURING THE GROWING PERIOD

Nikolina SHOPOVA[1], Dimka HAYTOVA[2]

Agricultural University, 4000 Plovdiv, 12 "Mendeleev" Str. Bulgaria, Email: nikpan@au-plovdiv.bg

Corresponding author email: nina_sm@abv.bg

Abstract

The aim of the study was to established the influence of age and the size of the planting area of the seedling plants grown in containers, on the vegetative behaviour of plants during the growing period in the conditions of late field production of tomatoes. The experiments were carried out during the period 2011 - 2013 on experimental field, Department of Horticulture at the Agriculturalal University – Plovdiv with cultivar 'Opal F1'. The variants with 20-25, 30-35 and 40-45 day seedlings, cultivated in containers with 40, 66 and 104 cells and planting area respectively 44, 28, 17cm^2, were tested. As a control was used 20-25 day seedling, grown in a transplanting bed and planting area 26-28 cm2 per plant (350-380 plant / m^2). It was found that the size of the planting area and the age of the seedling plants influenced on the vegetative behaviour of plants during the growing period after planting. The processes of growth and development of the plants, during the growth period, are most intensive at the variant with 20-25 days seedlings, grown in containers with 66 cells.

Key words: tomato seedlings, planting area, seedlings age, leaf area.

INTRODUCTION

The use of high quality seedlings is important to realize biological potential of plants in field production of tomatoes and obtaining better economic results (Kim Yeong-Bong et al., 1999; Singh et al., 2005).

Growing seedlings in containers is an advanced technological variant winning strong recognition in the field production of vegetables (Csizinszky and Schuster, 1988; Weston, 1988; Liu and Latimer, 1995).

For late field production of tomatoes the seedling production is done on transplanting bed. Cultivation of seedlings in containers is a progressive technological option.

In Bulgaria research on container-grown tomato seedlings is scanty (Simidchiev and Kanarska, 1986) and no studies are done on the late field production of tomatoes.

There is no scientific information on vegetative behavior of the plants during the growing season when using seedlings, grown in containers.

Unclear are a number of important questions relating to the parameters of the containers and the biology of the seedlings, and the influence of these factors on the growth and development of plants during the growing season.

The aim of the investigation is to determine the influence of the age of the seedlings and the size of the provided nutritional area on the vegetative behavior of plants during the growing period.

MATERIALS AND METHODS

The experimental work was carried out during the period 2011-2013 in an educational-experimental field of the Department of Horticulture.

For the cultivation of seedlings was used standard peat-pearlite mixture with components: peat by Durpeta of Lithuania and Agroperlit in a 3: 1 by volume. Peat substrate is pre-enriched N-250 mg / l, P_2O_5 - 270 mg / l, K_2O, 270 mg / l and Fe, Cu, Mn, B, Mo, Zn - 1,2 mg / l.

The salt concentration of peat, measured in ms / cm is 1.2, and the pH - 5.5-6.5.

The sowing of seeds was carried out in 3 types of containers made from expanded polystyrene

(EPS) with 40, 66 and 104-cells, providing nutritional area of a plant, respectively 44, 28 and 17 cm^2.

In the containers was grown seedlings of three ages - 40-45 days, 30-35 days and 20-25 days. To provide this age difference of the seedlings on the day of planting, the seeds were seeded in containers at an interval of 10 days and placed under optimum conditions for germination.

The control seedlings was grown in a transplanting bed, formed of the same mixture, at a rate of 2 g/m^2, the number of transplants being 350-380/m^2 by applying thinning during the second true leaf stage.

The sowing of seed was performed the period 30.05-04.06.

Setting up, the experiment was carrying out in the scheme of the block method in four replications.

The plant was plant in a permanent place in early July, in high bed-furrow surface in two-roll band in a schema - 110 + 50/30 cm. Growing them was done at the adopted technology for late field production with attachment on low wire construction with regular branches and one stem formation, with removal of vegetation peak after shaping fourth truss.

On the field the plant was planting during the period July 1-5.

The biometric measurements were performed on the 20th, 40th and 60th day after planting with parameters: height of the stem, number of leaves, leaf area and fresh weight of stem and leaves.

The leaf area was determined empirically by the formula (Konyaev M., 1970).

The mathematical processing of the data was done by using standard software SPSS – Duncan's Multiple Range Test (Duncan, 1955) and BIOSTAT.

RESULTS AND DISCUSSIONS

The results of the biometric measurement made on the 20th day after planting (Table 1) show differences in the values of the investigated biometric indicators. From the average results can be seen that for this stage of development of the plants the height of stem is greatest (53 cm) at 40-45 day seedlings, grown in containers with 40 cells, and the smallest (31.7)

at 20-25 day seedlings, grown in containers with 104 cells. The values of this indicator decrease by decreasing the age of seedlings and increasing the number of container cells.

The differences between all investigated variants for the index number of leaves are smaller.

Average for the experimental period, the formed leaf area is greatest at the variant of the third age group, grown in containers with 40 cells, which exceeded the control variant with 18.8%.

The leaf area of plants compared to the control is greater at the all variants with 66 cells - 4.9% to 7.3%. Only in the variants with 104 cells, reported values are smaller compared to the control.

Table 1. Biometric indicators of plants on the 20th day after planting average for the period

Number of cells (plant area, cm^2)	High of stem (cm)	Number of leaves	Leaf area		Fresh mass stem + leaves,	
			cm^2	% to the control	g	% to the control
40-45 day seedlings						
1. 40 (44)	53.0	13.5	1696.0	115.8	143.3	123.2
2. 66 (28)	46.0	12.0	1571.3	107.3	128.7	110.6
3.104 (17)	42.0	11.0	1274.7	87.0	108.7	93.4
30-35 day seedlings						
4. 40 (44)	48.7	12.2	1612.0	110.1	137.0	117.8
5. 66 (28)	41.7	11.4	1537.7	105.0	129.0	110.9
6. 104 (17)	35.3	11.1	1295.0	88.4	110.7	95.2
20-25 day seedlings						
7. 40 (44)	37.7	11.1	1740.0	118.8	147.0	12.4
8. 66 (28)	33.0	10.6	1536.0	104.9	121.0	104.0
9. 104 (17)	31.7	10.0	1283.0	87.6	105.3	90.6
10. control	39.3	10.2	1464.3	100.0	116.3	100.0

The basic indicator of the vegetative manifestations of plants is the size of the formed fresh vegetative mass.

For this indicator the highest average values during the experimental period were recorded in plants of the third (20-25) age group, grown in containers with 40 cells, followed by the same variant of the first (40-45) age group.

As can be seen from the results for these two variants the increase relative to the control is with 26.4% and 23.2%.

The reported values for the variant with 66 cells of the third age group are relatively high. They exceed the control with 4.0% to 10.9%. Only for variants with 104 cells the fresh phytomass is smaller compared to the control, as the average values are 4.8% - 9.4% lower.

The presented results and the fact that at the

time of planting the 40-45 day seedlings has a greatest leaf area and vegetative mass, may can considered that by the 20th day after planting the biologically younger plants are growing more intensively.

The growth rate of vegetative organs is greatest between the 20th and 40th day after planting, when plants form the main part of the leaves-stem mass. This biological feature is characteristic of the later field tomatoes (Cholakov, 1987) and is related to the fact that at the end of this period, the fruits formed of first and second truss are growing intensive, which requires the synthesis of larger quantities of organic substance.

At this measurement (Table 2), the reported differences between the variants by indicators of the stem height and number of leaves of one plant are smaller.

Average for the period the values for the stem height and number of leaves are lowest at variants of the third age group, in which falls and the control.

Table 2. Biometric indicators of plants on the 40th day after planting, average for the period

Number of cells (plant area, cm^2)	High of stem (cm)	Number of leaves	Leaf area		Fresh mass stem + leaves,	
			cm^2	% to the control	g	% to the control
40-45 day seedlings						
1. 40 (44)	107.7	21.6	5260.7	106.2	597.7	102.5
2. 66 (28)	102.3	20.6	5563.3	112.3	645.7	110.7
3.104 (17)	103.7	20.1	4876.3	98.4	580.3	99.5
30-35 day seedlings						
4. 40 (44)	106.3	20.9	5283.3	106.6	605.3	103.8
5. 66 (28)	102.0	20.3	5901.0	119.1	669.7	114.8
6. 104 (17)	101.3	19.8	5330.0	107.6	605.0	103.7
20-25 day seedlings						
7. 40 (44)	102.0	19.0	5944.0	120.0	665.3	114.1
8. 66 (28)	96.7	18.8	6199.3	125.1	695.3	119.2
9. 104 (17)	84.3	18.7	5688.0	114.8	639.7	109.7
10. control	96.7	19.1	4954.0	100.0	583.3	100.0

A comparison the variants on leaf area and stem + leaves mass show several trends.

On the 40th day after planting, the values for both indicators are highest at the plants, grown in containers with 66 cells.

For the leaf area the increase relative to control is respectively 12.3%, 19.1% and 25.1% and it's highest in the third age group.

Well expressed is the advantage of variants with 66 cells in terms of indicator fresh mass stem + leaves.

Average for the period of the study, the increase compared the control is respectively 10.7%, 14.8% and 19.2% and it's highest in the third age group.

Comparison of the variants with the same number of cells of the three age groups showed an advantage for the third (20-25), in which are recorded the highest values.

Between 40th and 60th day after planting the growth rate of vegetative organs reduced (Table 3).

In the middle of this period, in all variants are formed the fruits in the fourth truss, and the most of the synthesized plastic materials are used for their grown.

In addition at the end of this period is committed removing of vegetation peak of plants. This is the reason for minor differences between the variants of the stem height and number of leaves.

The plants, grown in containers with 66 cells stand out.

At that measurement, the values of leaf area and fresh vegetative mass in those variants are greatest. The increase, relative to control for the first indicator is respectively by 0,8%, 4.2% and 12.5%, and for the second - by 1.3%, 7.4% and 13.7%.The values for the both indicator are highest at the variant from the third age group.

Table 3. Biometric indicators of plants on the 60th day after planting, average for the period

Number of cells (plant area, cm^2)	High of stem (cm)	Number of leaves	Leaf area		Fresh mass stem + leaves	
			cm^2	% to the control	g	% to the control
40-45 day seedlings						
1. 40 (44)	113.7	21.5	6027.7	96.5	786.3	95.7
2. 66 (28)	110.0	21.2	6293.7	100.8	831.9	101.3
3.104 (17)	110.0	20.3	5674.7	90.9	750.0	91.3
30-35 day seedlings						
4. 40 (44)	105.0	20.7	6054.7	96.9	818.0	99.6
5. 66 (28)	104.3	20.8	6510.0	104.2	881.7	107.4
6. 104 (17)	107.3	19.7	6100.3	97.7	850.7	103.6
20-25 day seedlings						
7. 40 (44)	105.3	19.7	6547.0	104.8	868.7	105.8
8. 66 (28)	105.3	20.0	7025.0	112.5	934.0	113.7
9. 104 (17)	104.0	19.8	6673.0	106.8	887.3	108.0
10. control	104.7	19.0	6246.0	100.0	821.3	100.0

CONCLUSIONS

The size of nutritional area and the age of seedling, grown in containers for late field production of tomatoes influenced the vegetative behaviors of plants, during the vegetation period after planting.

On the 20th day after planting, the formed leaf area and fresh vegetative mass increases with decrease the number of container cells size. The reported values for both indicators are greater at the variant from third age group with 40 cells. The increases compared to the control are 118.8 % for leaf area and 126, 4 % for fresh vegetative mass.

The growth of leaf area and vegetative mass is greater between 20th and 40th day after planting. For the three age groups, reported values are greater at the variants with 66 cells.

At the end of the reporting period, the plants of variants with 66 cells of third age group form a large leaf area and fresh vegetative mass and increasing compared to the control is respectively with 12.5% and 13.7%.

REFERENCES

Cholakov D., 1987. A contribution to the study of determinants tomato varieties for late field production. Plant Science, item. XXIV, № 6, p. 64-68.

Csizinszky A.A., Schuster D.J., 1993. Impact of insecticide schedule, N and K rates, and transplant container size on cabbage yield. HortScience 28:299-301.

Duncan D., 1955. Multiply range and multiple F-tests. *Biometrics*, 11:1-42.

Kim Yeong-Bong, Hwang Yeon-Heon, Shin Won-Kyo, 1999. Effects of root container size and seedling age on growth and yield of tomato. Journal of the Korean Society for Horticultural Science 40(2): 163-165.

Konyaev N. F., 1970. Mathematical method for determining the area of plant leaves. – In: Reports VASHNIL, № 9, p. 5-9.

Liu A. and Latimer J.G., 1995. Root cell volume in the planter flat affects watermelon seedling development and fruit yield. HortScience 30:242-246.

Simidchiev, H., Kanazirska V., 1986. New technologies in Seedlings. In: M.Yordanov, (Editor) Advanced technologies in agriculture, pp. 150-180.

Singh B., Yadav H.L., Kumar M., Sirohi N.P.S., 2005. Effect of plastic plug-tray cell size and shape on quality of soilless media grown tomato seedlings . Acta Horticulture 742: International Conference and Exhibition on Soilless, Culture: ICESC.

Weston L. A., 1988. Effect of flat cell size transplant age and production site on growth and yield of pepper transplants. HortScience 22(4):709-711.

13

NEW GENOTYPES OF EGGPLANTS OBTAINED AT V.R.D.S. BUZĂU

Camelia BRATU[1], Florin STĂNICĂ[2], Costel VÎNĂTORU[1], Viorica LAGUNOVSCHI[2], Bianca ZAMFIR[1], Elena BĂRCANU[1]

[1]Vegetable Research and Development Station Buzău, No. 23, Mesteacănului Street, zip code 120024, Buzău, Romania

[2]University of Agronomic Sciences and Veterinary Medicine of Bucharest, 59 Marasti Blvd., District 1, Bucharest, Romania
Corresponding author email: botea_camelia2007@yahoo.com

Abstract

The V.R.D.S. Buzău breeding laboratory has put a great emphasis on maintaining the biodiversity of this species by constituting a valuable germplasm collection which has 84 genotypes and it is still growing. Our unit, V.R.D.S. Buzău, has patented so far two varieties, 'Dragaica' and 'Zaraza', and recently the F1 hybrid Rebeca. After evaluating their genetic stability it was found that 36 genotypes are stable, 23 are in an advanced form of stabilization and 25 accessions are still segregating. Researches completed until now by stabilizing an important number of valuable accessions with distinct phenotypic expressivity: A 10 with white fruits, A25A with brindle markings of white and purple, A 26 has purple fruits, A20 with small egg-shaped fruits, A29A. A29B, A29C and A29D with small red fruits at the physiological maturity, arranged in raceme, and of different shapes: round, ovoid, pumpkin etc. From all this varieties, the A10 accession was proposed for patenting and homologation, and will be followed by other varieties that are still under evaluation.

Key words: breeding, varieties, Rebeca F1.

INTRODUCTION

Our country has favorable pedo-climatic conditions for cultivating the *Solanum melongena* species. Preoccupations in breeding this species have been since V.R.D.S. Buzău was founded. For starters, imported varieties were cultivated, and in time, the researchers from V.R.D.S. Buzău achieved valuable genotypes that were very appreciated both by growers and consumers, among which the variety 'Dragaica', destined for protected areas and open field cultivation, the 'Zaraza' variety destined only for open field cultivation and, recently the F1 hybrid 'Rebeca' has been obtained, with a mixed destination, protected areas and open field culture.
Researches were constituted among achieving a rich germplasm collection for this species, comprising local populations, accessions, autochthonous and foreign varieties. "Most species within *Solanum* are endemic to the Americas; however, ~20% is Old World species. The common name eggplant encompasses three closely related cultivated species that belong to *Solanum* subgenus

Leptostemonum: *Solanum melongena* L., brinjal eggplant or aubergine; *S. aethiopicum* L., scarlet eggplant; *S. macrocarpon* L., gboma eggplant." (Daunay et al., 2001).
Among the old and traditional cultivars, an emphasis was put on preserving this species biodiversity.
"The first center of diversity for eggplants is in India, and the second one in China." (Ramalho do Rêgo, 2012)
Thorough specific breeding work programs we can achieve new distinct genotypes.
"Eggplants (*Solanum melongena* L.) were domesticated in tropical Asia where they are used abundantly as both food and medicine. Human selection has produced hundreds of landraces that differ in morphology and chemistry in ways that may be related to local ethnobotanical preferences." (Meyer et al., 2012).

MATERIALS AND METHODS

Researches were targeted on collecting valuable genetic material and structuring it based on the breeding objectives. The general field collection has a large number of genotypes that are in different stages of breeding, among which 84 accessions were promoted in the working field and submitted to an intensive breeding program. Among these, 36 accessions are genetically stabilized, 23 are in an advanced stage and 25 are still segregating. From all the 36 stabilized accessions, 11 part of this paper due to their distinct phenotypic expressivity.

The breeding methods used were repeated individual selection, hybridization, segregation and negative mass selection. The crop technology used both for protected areas and open field was the species specific one. Crop establishment was made through seedlings aged of 55-60 days.

For the open field crop, the planting scheme used was of 70 cm between rows and 35-40 cm between plants on rows, and for the protected areas, the crop establishment was made in bands at 70 cm between rows, 1.2 m between bands and 40 cm between plants on the rows (Figures 1 and 2).

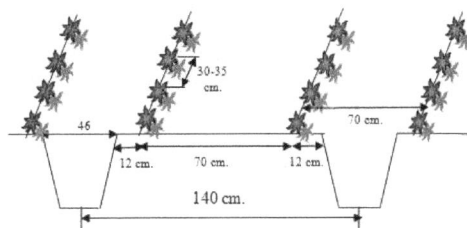

Figure1. Planting scheme for protected areas

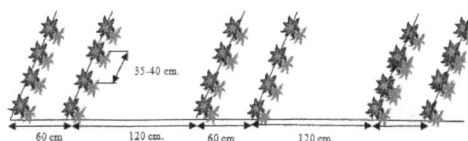

Figure2. Planting scheme for open field

RESULTS AND DISCUSSIONS

The research conducted since 1990 at V.R.D.S Buzău, finalized with the achievement of ten genetically stabilized accessions and with distinct phenotypic expressivity: A 1 S, A 7, A 10, A 20, A 25 A, A 22 A, A 23 A, A 24 A, A 29 A, A 29 B.

The main plant characteristics are presented in table 1.

The values recorded in table 1 demonstrates visible distictibility between the accessions in what regards the main traits of the eggplants. Regarding the plant height, on the first place is A 29 B, with a medium height of 180 cm and

the lowest value was recorded at L 20 with 112 cm height. Differences were recorded regarding thr anthocianic coloration. It was founded that the accessions A 1 S, A 22 A and A 23 A have a strong anthocianic pigmentation on the stem, sprouts and on the main leaf veins. Due to this characteristic these eggplants weren't preffered by the Colorado beetle.

The foliage is an important marker for distictivness for all the new creations. Important differences were observed regarding the length of the stalk, the length and width of the lamina, and leaf type.

Table 1. The main characteristics of the plants, mean values

Character/accession	Rebeca F1	L1 S	L7	L 10	L 20	L 25 A	L 22 A	L 23 A	L 24 A	L 29 A	L 29 B
Plant height(cm)	155	147	125	120	112	129	128	118	126	156	180
Stem height (cm)	14	26	13	9	13	22	33	19	47	8	19
Main sprouts no.	3	4	2	5	2	2	2	4	4	3	4
Main sprouts length (cm)	116	123	73	91	99	83	87	98	71	115	103
Stem diameter (cm)	2.2	2.3	1.5	2.3	1,6	1.4	1.9	1.7	1.5	2.1	1.8
Peduncle length (cm)	13	11	11	11	11.5	15	8	8.5	11	10.5	12
Lamina length (cm)	27	21	14.5	29	25	35	29.5	30.5	33	33.5	31
Lamina width (cm)	19	18	17.5	22	15.5	20	20	18.5	15.5	22	27
Flower color	Mauve	Mauve	Purple	Purple white	Mauve	Purple white	Purple	Mauve	Mauve	White	White
Corolla diameter (mm)	35	32	48	45	31	42	35	40	53	24	20
Anthocianic pigmentation	-	Stem, sprouts, leaf, sepals	-	-	-	-	Sprouts, main and sec. veins	Stem, main veins	-	-	-

A special attention for fruit setting was given in the breeding. The accessions achieved are distinguished by productivity, quality, earliness and genetic resistance to the main specific diseases. The researches conducted had as the main purpose, the assortment enrichment with new cultivars at this species, valuing its genetic potential by introducing new totally different genotypes beside the old, classic ones.

A special emphasis, was on the color and shape of the fruit, correlated with the directions of use (Figure 3.).

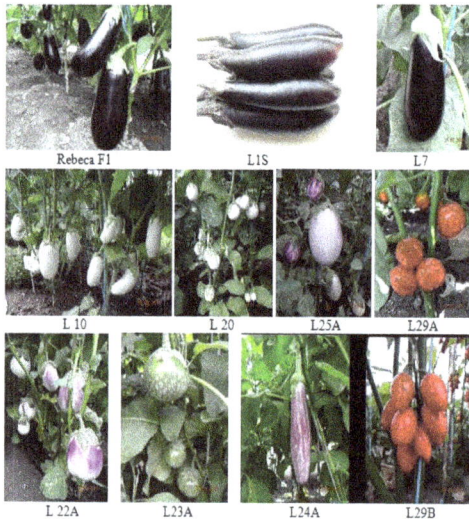

Figure 3. Main accessions studied

The fruit main traits are presented in table 2. Throughout the experience, V.R.D.S. Buzău managed to obtain the F1 hybrid Rebeca, a hybrid that can be used in protected areas, but also in a open field crop.

It is presented as a globular bush, with an average height of 155 cm and has no spines. The color of the fruit is black and shiny, cilindrically shaped.

It preserves the characteristics of the traditional eggplants that have a large fruit. The average fruit weight is of 590 g, has a small number of seeds that are mostly present in the apex of the fruit. This variety is genetically resistent at the main pathogens. Average production per plant is registered with 3.5 kg.

A 1 S has long black fruits, sword-type, with an average length of 26 cm and 8.4 cm median diameter. This accession is distinguished by a high productivity, strength, the presence of anthocianic pigmentation on the plants vegetative organs, including the fruit sepals. Due to it's strength and strong radicular system, well developed, it can be recomended to be used as a rootstock. This genotype has the "*mi*" gene, known as a repelent gene for nematodes. The average plant production is of 2.8 kg.

A 7 has medium length fruits, with green sepals, a small number of seeds dispersed throughout the length of the fruit. The average fruit weight is of 457.4 g, and 2.7 kg fruit/plant production.

A 10 has large white fruits, a trait that distinguishes it from the other varieties, along with a reduced number of seeds. It is highly productive, with fruits that have a good quality, and the average production per plant   is of 4.1 kg. Due to it's exceptional traits, productivity, earliness, along with fruit quality and good process capacity made this genotype to be highly demanded by growers and consumers, and the request for seeds and seedlings grew significantly year by year.

A 20 is a premier for our country, especially for culinar dishes; it is distinguished by its small white, egg-shaped fruits. It can be used in a great variety of dishes in order to replace mushrooms, or can be pickeled as cucumbers and mushrooms. It has a large variety of fruit uses and it can also be used as in an ornamental system. The fruit has an average weight of 49.2 g and the registred production/plant is of 1.7 kg.

A 25 A is distinguished by white fruits that have purple shadows, and a mean fruit weight of 155.5 g with an average production/plant of 1.3 kg. Regarding the cpacity of production is below the other genotypes, but, due to its shape and colour it is highly demanded both by consumers and growers.

Another trait of distinctibility are the spines that are very much present on the sepals and peduncle, giving it a rustic aproach. Also, it was recorded an increased resistance to specific diseases, especially to *Verticilium*.

A 22 A has large, obovate, white with purple patches fruits. They are over 600 g, and as a special trait has ribbs. This cultivar is remarkable by its productivity, obtaining the highest production/plant, 4.7 kg. All of this features made this genotype to have a great request from growers and consumers, for open field crops and for protected areas.

A 23 A, a cultivar remarked by the shape and colour of the fruits, white with green patches, small rounded shape.

The average weight is 101.6 g and the mean production/plant is of 1.3 kg. Its shape and colour makes it very special, and we believe it will have great opportunities for the future.

A 24 A is distinguished by its productivity, quality, and the main feature is represented by the fruit colour. It has stripped fruits, white with mauve stripes. The fruit average weight is 530 g and the total production per plant is of 3.7 kg. Due to its attractive shape and colour that gives it a special appearence, this genotype is more and more demanded by the market.

Table 2. Fruit characteristics

Character/accession		Rebeca F1	L1 S	L7	L 10	L 20	L 25 A	L 22 A	L 23 A	L 24 A	L 29 A	L 29 B
Peduncle length (cm)		12	5.9	7.3	9,8	2.9	6.6	5	5.6	6.3	1.8	1,6
Spines presence on		Sepals	-	Sepals	Peduncle	Sepals	Sepals, peduncle	Sepals	-	Peduncle, sepals	-	-
Sepals colour		Green	Green	Green	Light green	Light green	Green	Green	Light green	Light green	Dark green	Green
Fruit length (cm)		24.3	26	18.5	18	5.4	12.6	10.5	5.2	14	5.5	4.3
Fruit diameter (cm)	Apex	6	4.3	7,1	7.5	3.6	6.1	10.7	5.2	7.9	5.7	1.6
	Median	8.4	5.7	6,4	9.1	4.4	8.5	12.5	6.3	9.6	6.3	2.4
	Base	5.4	4.4	4,5	7	4.3	6.9	10.7	5.2	7.2	5.7	1.8
Fruit colour	Technological maturity	Black	Black	Black-mauve	White	White	White + mauve	White + mauve (patches)	White/green patch	White+ mauve patches	Stripped green	Beige+ green
	Physiological maturity	Black	Black	Dark mauve-dark green	Yellow	Yellow	Yellow+ mauve	Yellow with mauve patches	Yellow	Yellow+ mauve	Red with green stripes	Dark orange
No of. fruits/plant		6	7	6	7	36	9	7	13	7	16	81
Fruit weight (g)		590	412.6	457.4	592	49.2	155.5	685	101.6	530	106.6	11.4
Total fruit weight/pl. (kg)		3.5	2.8	2.7	4.1	1.7	1.3	4.7	1.3	3.7	1.6	0.92

A 29 A is a new valuable genotype obtained at V.R.D.S. Buzău, that is distinguished by green fruits that colours in red at the physiological maturity, with small flat-globular fruits. The fruit has an average weight of 106.6 g and a total production/plant of 1.6 kg. This cultivar is characterized by a great vigour, making it suitable for the ongoing tests as a rootstock. So far, the preliminary results showed a great compatibility between it and the scions used, imprinting them a great resistance to the specific pests and pathogens.

A 29 B is a special variety that is distinguished by its pepper-type flower, and and linear raceme type inflorescence. The fruits are small, ovoid, beige with small green stripes. The average fruit weight is of 11.4 g and total production/plant is 0.9 kg. This cultivar is also being tested as a rootstock, with very good results.

CONCLUSIONS

Researches finalized by setting a solid germplasm collection, evaluation and computerized data gathering of new stable in lineage genotypes. From all of them, due to its special appearance and features, A 10 was in 2017 registered at ISTIS Bucharest to be patented under provisional name 'Camelia'.

The achieved genetic resources allow us to obtain new valuable cultivars.

REFERENCES

Doganlar S., Frary A, Daunay MC, Lester RN, Tanksley SD., 2002. A comparative genetic linkage map of eggplant (Solanum melongena) and its implications for genome evolution in the Solanaceae, Genetics 161(4): 1697-1711.

Meyer R.S., 2012. Chemical, genetic, and ethnobotanical diversity in Asian eggplant. City University of New York.

Ramalho do Rêgo E., Nascimento M.F., do Nascimento N.F.F., Fortunato F.L.G., Finger F.L., Monteiro do Rêgo M., 2013. Heterosis for fruit quality traits in ornamental peppers, Breakthroughs in the Genetics and Breeding of Capsicum and Eggplant, Torino, 2-4.09.2013; 701

14

DETECTION AND IDENTIFICATION OF ALTERNARIA SPECIES CAUSING DISEASES OF CARROT IN ANKARA PROVINCE, TURKEY

Senem TÜLEK[1], Fatma Sara DOLAR[2]

[1]Ministry of Agriculture and Rural Affairs, Central Plant Protection Research Institute, 06172, Yenimahalle, Ankara, Turkey, Email: senemtulek@gmail.com

[2]Ankara University, Faculty of Agriculture, Department of Plant Protection, 06110, Ankara, Turkey, Email: fsdolar@gmail.com
Corresponding author email: fsdolar@gmail.com

Abstract

Carrot (Daucus carota var. sativus) is widely planted in the Ankara provinces. In order to identify species of Alternaria causing disease on root and foliage, surveys of carrot production areas in Ayaş and Beypazarı districts of Ankara province between June and November in 2008–2009 were undertaken. Sixty isolates of Alternaria spp. have been obtained from necrotic lesions on the leaves and roots. Alternaria radicina, A. alternata, A.tenuissima and A.dauci were isolated from symptomatic plants collected in our survey and the pathogenicity of fungi have been tested. Species identification was done based on culture and conidial morphology, growth rate and rDNA sequences. Pathogenicity test such as hypocotly test, carrot disc method, 6–8 weekly seedling andplant test were conducted Isolates of A.radicina, A.dauci were shown high virulence although Alternaria alternata were found as moderately or low virulence.

Key words: Carrot, root rot, Alternaria blight, Alternaria spp., Turkey.

INTRODUCTION

Carrot (*Daucus carota* **var.** *sativus* Röhl.) is one of the most popular and commonly consumed vegetables (Rubatsky, 2002). Commercial carrot production is an economically important industry worldwide. In the Turkey, the most productive land is in the Central Anatolia Region and Eastern Mediterranean Region. Carrot is widely planted in the Anatolia region that includes Ankara producing nearly 60% of Turkey's carrots. In 2013, carrot cultivated area is about 104,404 da and annual production is 557,977 tons in Turkey (Anonymous, 2014). Root and foliar diseases are among the most important factors limiting carrot production worldwide.

Fungi are the most common pathogens of *D. carota*. Species of the genus *Alternaria* such as *Alternaria carotiincultae* E. G. Simmons, *Alternaria dauci* (J. G. Kuhn) J. W. Groves & Skolko, *Alternaria. petroselini* (Neerg.) E. G. Simmons, and *Alternaria radicina* Meir, Drechsler &E. D. Eddy, have been reported on *D. carota* for several countries (Farrar et al., 2004).

Alternaria Nees is an widely spread mould genus which can be found on plants, in soil, food and indoor air. Most frequent species are *A. alternata* and *A. tenuissima* (Lõiveke et al., 2004). The pathogenic species *Alternaria radicina* and *Alternaria dauci* are isolated from diseased carrot plants in all growing stages (Stranberg, 2002). Alternaria leaf blight caused by *A. dauci* and Alternaria black rot caused by *A. radicina* are widespread on carrot crops in the world where are reported to cause considerable damage (Davis and Raid, 2002).

Black rot (*Alternaria radicina*) is found in all the main carrot-production areas. Although this disease is important as a storage disease of carrots, it also causes seedling damping-off, foliar and crown infection ((Koike et al., 2009). Alternaria leaf blight (*Alternaria dauci*) is one of the most important foliar diseases of carrot and occurs worldwide. Severe epidemics reduce carrot root size and yields (Koike et al., 2009). In Turkey, *A.dauci* was first described as leaf blight caused on carrot in the Hatay province of Turkey (Soylu et al., 2005). The objective of the present study was to determine the *Alternaria* species causing diseases in carrot growing areas in Ankara province, Turkey.

MATERIALS AND METHODS

Survey and fungal isolation

In order to identify species of *Alternaria* causing diseases on carrot root and foliage, surveys were carried out in production areas in Ankara province between June and November in 2008-2009 growing seasons (Figure 1). Samples were taken from fields in Ayaş and Beypazarı districts of Ankara.

Infected carrot leaves and root pieces were surface-sterilized (1,0% (w/v) sodium hypochlorite) for 2-3 min then rinsed in sterile water three times before they were placed onto potato dextrose agar (PDA, Merck) containing streptomycin and incubated at 23±1°C with a 12-h photoperiod for 7-10 days. Single spore isolates were stored on PDA slant tubes at 4°C.

Figure 1. Survey area for Alternaria diseases of carrot in Turkey

Identification of fungus

For identification, culture morphology, growth rate and conidial morphology were observed from 12-15 day-old cultures grown on PDA and PCA (Ellis 1970, 1971; Rotem 1994; Simmons 1995). The shape, length and width of 50 conidia for each isolate were determined and mean length and width were calculated. In addition, the number of transepta per conidium and the production of conidia in catenate arrangement was determined.

Pathogenicity tests

The pathogenicity of isolated fungi from diseased plants was assessed. Carrot disc pathogenicity test (modified from Pryor et al., 1994), the 6-8 week old seedling test (Coles and Wicks 2003) and plant test (Pryor and Gilbertson 2002) were used.

Carrot disc pathogenicity tests

Mature "Maestro" carrots were assessed for pathogenicity of *Alternaria* spp. Mature carrot roots were washed in tap water and sliced into disks approximately 5 mm thick. The disks were surface-disinfested by soaking in 0,1% sodium hypochlorite for 5 min. then triple rinsed with water and placed on a paper towel for 1hr to dry. The four carrot discs were then placed in each petri dishes (20 x 100 mm) containing two Whatman No. 1 filter papers moistened with 2 ml of streptomycin sulphate solution (100mg/l) Twenty discs were used for each isolate. Carrot discs were inoculated with mycelial plugs (4 mm diameter) cut from the margins of actively growing culture (Figure 2). Controls were treated similarly using similar sized pieces of water agar. The dishes containing inoculated disks were incubated on wire racks in clear plastic trays for 10 days at 24 ± 2 ° C, 12 h light with 12-hour dark cycle. After 10 days, pathogenicity was evaluated on a scale of 0 to 4 (Coles and Wicks, 2003). A total of 60 isolates of Alternaria. was tested, and each test was replicated four times.

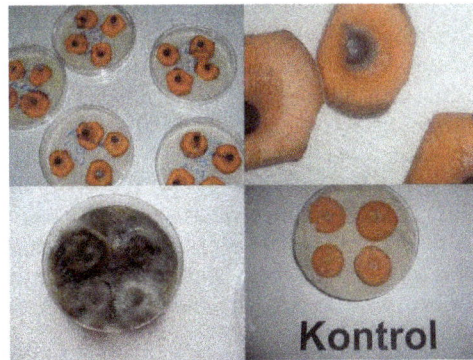

Figure 2. Carrot disc pathogenicity tests

Six-eight week old seedling test

The second method used fresh 6 and 8-week old carrot seedlings of the commercial cultivar "Nantes". Seedlings were placed on surface sterilised aluminium foil sheets in prewashed plastic trays with pre-moistened absorbent paper. Five seedlings per treatment were inoculated by taking 1.0 x 0.5 cm water agar pieces from mature colonies of *A. radicina* and placing the mycelial surface down on to the hypocotyl region near the crown of each seedling (Figure 3). Controls were treated similarly using similar sized pieces of water agar. The trays were enclosed in a clear plastic bag and incubated on the laboratory bench at room temperature for 10 days. The level of disease was assessed by measuring the extent of necrosis from the point of inoculum. Fungi causing blackening, or soft decomposition of the hypocotyl region, or death of the upper stem and petioles were classed as pathogenic. After 10 days, fungal growth and pathogenicity were evaluated on each plant based on 0-4 scale (Coles and Wicks 2003). Each treatment was replicated three times.

Figure 3. Six-eight week old seedling test

Plant test

Eight seeds from Nantes variety were sown in 10.5 cm diameter plastic pots containing a sterilized mixture of carrot field soil, peat and sand (1:1:1,v/v/v). Pots were maintained under optimum greenhouse conditions at temperatures ranging from 23–26 °C, and 35–40% humidity.

Conidial suspensions were prepared in sterile distilled water using 14-day-old cultures. Spore suspensions were adjusted to $2x10^3$ conidia/mL for *A. radicina* (Pryor and Gilbertson, 2002) and other *Alternaria* spp. $1x10^3$ (Pryor et al., 2002) and sprayed onto areial parts of each test plant, until run-off, with an pressure hand sprayer. Controls were sprayed with sterile distilled water. Four replicates were used for each isolates.

Two weeks after inoculation, pathogenicity were evaluated on 0 to 5 scale (Pryor and Gilbertson, 2002).

Disease assessment

Isolates of *Alternaria* spp. were assessed for their pathogenicity on carrot disc and 6-8 week old seedling test using 0 to 4 scale from Coles and Wicks (2003): 0= no discoloration, 1=slight discoloration, 2= slight discoloration with mycelial growth, 3= grey to black necrosis with the production of conidia, 4 = grey to black necrosis with abundant production of conidia.

Isolates of *A. radicina* and *A. dauci* were assessed for their pathogenicity on plant test using a 0 to 5 scale from Pryor and Gilbertson (2002):
0 = no disease, 1= 1% leaf necrosis, 2= 5% leaf necrosis, 3= 10% leaf necrosis, 4= 20% leaf necrosis, 5= more than 40% leaf necrosis

These scale values were converted to disease severity values (Xi et al., 1990) using the following formula:

$$\text{Disease sev.}= \frac{\Sigma(\text{no. of plant in category x category value}) \times 100}{\text{max. category value x total no.of plants}}$$

The isolates were classified according to disease severity values such as highly virulent (75-100%), moderately virulent (50-74,9 %) and weakly virulent 0-49,9%).

The data were subjected to ANOVA, and the means were separated by the least significant difference (LSD) test.

Molecular analysis

Approximately, 300 mg mycelium were harvested and ground with liquid nitrogen in a sterile mortar for DNA extraction from culture medium. Genomic DNA was extracted using a Qiagen DNeasy ®Plant Mini Kit, as specified by the manufacturer, and stored at 20 °C prior to use. PCR reaction mixtures and condition were modified from previous studies (Aroca and Raposo 2007; Cobos and Martin, 2008). The reaction mixtures of PCR, a final volume of 50 µl, contained 5µl of 10X buffer [75 mM Tris HCl, pH 9.0, 50 mM KCl, 20 mM (NH4)2SO4], 2 µl of 5 µM each primers, 5 µl of 1.5mM MgCl2, 2 µl of 10 mM deoxynucleoside triphosphates (dNTPs), 1 U Taq polymerase (Fermatas), 5 µl of DNA template for each reaction and 5 µl of bovine serum albumin (BSA: 10 mg/ml).

DNA amplifications were carried out in a Techne TC-5000 thermal cycler by the following program: 94 C for 2 min, followed by 34 cycles of (1) denaturation (94°C for 30 s), (2) annealing (60 °C for 30 s) and (3) extention (72 °C for 30 s), and a final extension step 10 min at 72 °C. The ITS region of the isolates was amplified using the universal primers ITS-1 (5' TCC GTA GGTGAA CCT GCGG 3') and ITS -4 (5'TCC TCC GCT TAT TGA TATGC3'). The PCR products were separated in 1.5 % agarose gels stained with ethidium bromide, and visualized under UV light. They were sequenced by REFGEN (Gene Research and Biotechnology Company, Ankara, Turkey).

RESULTS AND DISCUSSIONS

Identification of Alternaria isolates and their pathogenicity

A total of 2,297 da carrot growing areas were surveyed in Ayaş and Beypazarı districts of Ankara province in 2008–2009. Sixty isolates of *Alternaria* were obtained from infected carrot root and foliage. *Alternaria radicina, A. alternata A. tenuissima, A. dauci,* were isolated from diseased plants collected in the survey. Of the identified isolates, 22,42% were *A. radicina,* 56,14% were *A. alternata,* 7,14 % were *A. tenuissima* and 14,28 % were *A. dauci.*

Our survey showed that *A. radicina* was associated with root and leaf of carrot and was widespread in carrot plantings in Ankara. The fungus was encountered most frequent from carrot rot and crown in summer and autumn. In our study *A. alternata* was obtained from dissected diseased tissue. *A. alternata* is one of the most common saprotrophs or facultative parasites associated with various parts of plants (Scheffer, 1992). Up to 68% of carrot root samples collected in several European countries were found to be contaminated with the fungus (Solfrizzo et al., 2005). As much as 70% of mature carrots can be rendered unmarketable if heavily infested or infected by *A. radicina* and *A. alternata* (Solfrizzo et al., 2005).

Fungal identification was confirmed by DNA sequencing.

Alternaria radicina

Alternaria radicina was isolated from roots and crown of young and mature carrots. Symptoms of the disease as the black rot was observed by dry, black, decay, sunken lesions on carrot roots. Lesions were quickly expand, and decay the entire root (Figure 4). Symptoms seen on the roots and crown of carrot seedlings were observed initially as small chlorotic spots and these spots were joined together by expanding. Lesioned tissues were significantly separated from the healthy tissue.

Figure 4. Black rot symptoms on the root and crown of carrot

During the survey it was observed that maturing carrots were often damaged around root regions. Our results showed that this was the effect of *Alternaria radicina* infection.

The colony color was dark green–blackish on PDA in 10-14 days. We shown that conidia were borne singly, or occasionally in chains of two, and were typically dark olive-brown to natal brown, broadly ellipsoid to ovoid, 12–17x19–37 µm, with one to four transepta and one to two longisepta in any or all segments, except basal and apical segments, which usually are free of septa (Figure 5).

Morphological features of our tested isolates on PDA were similar with descriptions of Ellis (1970, 1971), Rotem (1994) and Simmons (1995).

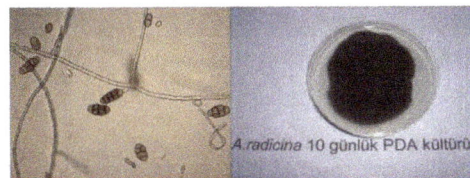

Figure 5. Morphology of conidia of *A. radicina* (x40): colony appearance of *A.radicina* on PDA

The resulting sequences were compared to other *A.radicina* sequences in the GenBank and were 99 and 100 % identical.

Isolates of *A.radicina* were tested using all three pathogenicity tests. All isolates showed a large variation in virulence. In the results of carrot disc pathogenicity tests, disease severity values of isolates were found between 52,1 to 75,0 %. Disease values were determined as 41,7-85,0% and 45,0-98,6% in seedling and plant tests, respectively. *A.radicina* caused the high and moderate disease ratio on carrot disc, seedling and plant in pathogenicity tests.

Alternaria alternata

Alternaria alternata was usually the fungus most frequently isolated from symptomatic root rot.

Colonies were usually black or olivaceous black on PDA. Conidiophores arising singly or in small groups, simple or branched, straight or flexuous, pale to mid olvaceous or brown, one or several conidial scars. Conidia were formed in long chains, obclavate, obpyriform, ovoid or elipsoidal, with up to 3-5 transverse and several longitudinal septa, overall length 9-11x20-32 μm and 5-16 chains (Figure 6).

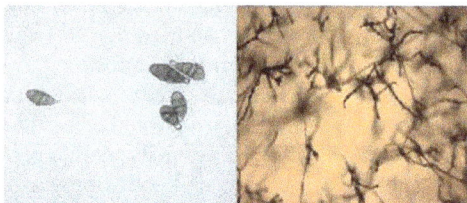

Figure 6. Morphology of conidia of *A.alternata* (x40) and chain structure (x20)

The resulting sequences were compared to other *A.alternaria* sequences and were 98-99% identical to other *A.alternata* sequences in the GenBank.

As a result of the pathogenicity test, we have found differences in virulence of tested isolates of *A. alternata.* Disease severity values of *A. alternata* were between 32,6 to 81,25% in carrot disc pathogenicity method.

Alternaria tenuissima

The fungus was isolated from symptomatic chlorotic leaf spot, discoloration and crown rot.

Colonies usually were pale black or olivaceous black on PDA. Conidiophores solitary or in groups, simple or branched, straight or flexuous, septate, pale brown, with one or several conidial scars.

Conidia formed 3-5 chains, obclavate, obpyriform or elipsoidal, generally with 3–5 transverse and several longitudinal, overall length 8–10x18–20 μm, beak measurment 5–9 μm (Figure 7).

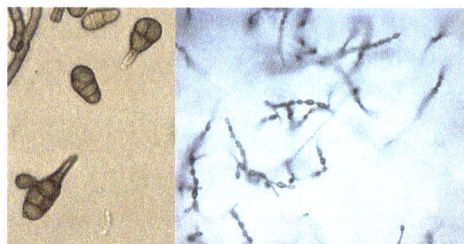

Figure 7. Morphology of conidia (x40) and chain structure of *A. tenuissima* (x 20)

The resulting sequences were compared to other *A. tenuissima* sequences in the GenBank and were 97 and 99% identical.

As a result of the carrot disc pathogenicity test, *A. tenuissima* isolates were found to be weakly pathogenic (31,3-35,4% disease severity) on carrot plants.

Alternaria dauci

During survey, foliage infection by *Alternaria dauci* was observed on carrot plants growing in Ankara. Initial symptoms first appeared on older leaves as irregularly-shaped, minute, dark brown-to-black spots, with yellow borders on the edge of the leaflet blade. As the disease progressed the lesions expanded, causing the leaflets to turn brown and die (Figure 8).

The fungus was consistently isolated from the margins of these lesions.

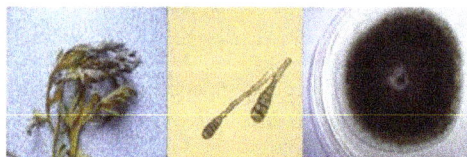

Figure 8. Alternaria leaf blight symptoms, morphology of conidia (x40) and colony appearance of *A.dauci* on PDA

The colony color was pale or dark gren-blackish on PDA in 10-14 days (Figure 8). Conidiophores were medium olivaceous brown, and either simple, with a single terminal conidiogenous site, or 1-2 geniculate and conidiogenous. Conidia were typically borne singly, but occasionall a sturdy terminal secondary conidiophore bearing a secondary spore is produced. Conidia were medium to dark olive-brown, long ellipsoid to obclavate, 10-22x45–70 μm (spore body), with 3 to 7 transepta and 1 to 2 longisepta in fewer than half of the transverse segments (Figure, 8). Mature conidia are rostrate with a terminal filamentous beak 30–120 μm, conidia occasionally in chains of single or two.

Morphological features of isolates on PDA were similar with descriptions of Ellis (1970, 1971), Rotem (1994) and Simmons (1995).

As a result of the plant pathogenicity test, we have found that *A.dauci* isolates were highly virulent with 89,25 to 92,75 % disease severity values.

CONCLUSIONS

Detection and identification of *Alternaria* species pathogenic in carrot growing areas in Ankara is fundamental to guide the development of appropriate strategies for disease management. *Alternaria radicina, A. alternata A. tenuissima, A. dauci* were identified through classical and molecular methods among the 60 isolates obtained from carrot growing areas in Ankara province.

Isolates of *A. radicina* and *A. dauci* showed high virulence although *A. tenuissima* were found as low virulent. *A. alternata* isolates were determined as moderately virulent. It was found differences among the virulence of isolates of *A. alternata*.

These results will be useful in developing of integrated strategies for disease management and breeding programs to Alternaria leaf blight and black rot disease on carrot.

ACKNOWLEDGEMENTS

We are grateful to Filiz ÜNAL because of molecular analysis of Alternaria isolates. This research was carried out with the support of Republic of Turkey Ministry of Food, Agriculture and Livestoc (TAGEM-BS-/08/10-09/02-07).

REFERENCES

Anonymous, 2014. Türkiye İstatistik Kurumu. www.tuik.gov.tr

Aroca, A. and Raposo, R., 2007. PCR-based stragety to detect and identify species of Phaeoacremonium causing grapevine diseases. *Applied and Environmental Microbiology*, 73, 2911-2918.

Cobos, R., Martin, M.T., 2008. Molecular characterisation of *Phaeomoniella chlamydospora*

isolated from grapevines in Castilla y León (Spain). *Phytopathol. Mediterr.*, 47, 20–27.

Coles, R. B. and Wicks, T. J. 2003.The incidence of *Alternaria radicina* on carrot seeds, seedlings and roots in South Australia Australasian Plant Pathology 32 (1), p. 99 – 104.

Davis, R. M. and Raid, R. N., 2002. Crown, Root, and Wilt Diseases. Compendium of Umbelliferous Crop Diseases, Pp : 25 - 40.

Ellis, M. B., 1970. More Dematious Hypomycetes. Commenwealth Mycol. England, 507.

Ellis, M. B., 1971. Dematious Hypomycetes. Commenwealth Mycol. England, 608.

Farrar, J. J., Pryor, B.M. and Davis, R. M., 2004. *Alternaria* diseases of carrot. Plant Disease, Vol 88; 776 - 784.

Koike S.T., Gladders, P. and Paulus, A.O., 2009. Vegetable Diseases, A Color Handbook. Academic Press.Third edition,Boston, 448.

Lõiveke, H., Ilumäe, E. and Laitamm, H., 2004. Microfungi in grain and grain feeds and their potential toxicity. Agronomy Research 2(2), 195–205. (in Estonian)

Pryor, B. M., Davis, R. M. and Gilbertson, R. L., 1994. Detection and eradication of *Alternaria radicina* on carrot seed. Plant Dis., 82: 891–895.

Pryor, B. M., and Gilbertson, R. L., 2002. Relationships and taxonomic status of *Alternaria radicina, A. carotiincultae*, and *A. petroselini* based upon morphological, biochemical, and molecular characteristics. Mycologia, 94 (1), 49–61.

Pryor, B. M., Strandberg, J. O., Davis, R. M., Nunez, J. J., and Gilbertson, R. L., 2002. Survival and persistence of *Alternaria dauci* in carrot cropping systems. Plant Dis. 86: 1115-1122.

Rotem, J., 1994. The genus *Alternaria*: biology, epidemiology and pathogenicity. The American Phytopathological Society, 326 pp.

Scheffer, R. P., 1992. Ecological and evolutionary roles of toxins from *Alternaria* species pathogenic to plants. In: Chełkowski J, and Visconti A [eds], *Alternaria*: Biology, plant diseases and metabolites, 101– 122.

Simmons, E. G., 1995. *Alternaria* themes and variations (112-144). Mycotaxon 55: 55-163.

Solfrizzo, M., Girolamo, A. De., Vitti, C., Tylkowska, K., Grabarkiewicz- Szczesna, J., Szopińska, D. and Dorna, H., 2005. Toxigenic profile of *Alternaria alternata* and *Alternaria radicina* occurring on umbelliferous plants. Food Additives and Contaminants, Volume 22, Number 4, p. 302–308.

Soylu, S., Kurt, S., Soylu, E. M., and Tok, F. M., 2005. First report of Alternaria leaf blight caused by *Alternaria dauci* on carrot in Turkey. Plant Pathology, 54; 252.

Strandberg, J. O., 2002. A Selective Medium for the Detection of *Alternaria dauci* and *Alternaria radicina*. Phytoparasitica, 30 (3) ; 269–284.

Xi K, Morrall RAA, Baker RJ and Verma PR., 1990. Relationship between incidence and severity of blackleg disease of rapeseed. Canadian Journal of Plant Pathology, 12:164-169.

15

CURRENT STRATEGIES FOR THE PROTECTION OF ORGANIC CROPS IN VEGETABLES PRODUCTION

Steliana RODINO[1], Marian BUTU[1], Gina FIDLER[1], Ancuța MARIN[2], Alina BUTU[1]

[1]National Institute of Research and Development for Biological Sciences, Splaiul Independentei, no 296, P.O. Box no 17-16, 060031, Bucharest, Romania

[2]Research Institute for Agricultural Economics and Rural Development, 61 Blvd. Marasti, 011464, Bucharest, Romania

Corresponding author email: marian.butu@yahoo.com

Abstract

Development of disease resistance to conventional pesticides and environmental contamination problems created pressure on farmers to adopt new strategies for disease control in the production of vegetables. In addition, consumers demand to minimize pesticide residues in food products is forcing the growers and the pesticide producing companies to formulate and develop alternatives to the contentious inputs currently marketed. Products obtained from biologically active compounds extracted from plants will play an increasingly important role in crop protection strategies. Exploiting antimicrobial substances from plants that inhibit or halt the reproduction of pathogenic microbes, would become a more realistic and ecological method for the integrated management of plant diseases with the final goal of reducing or gradually phasing out contentious inputs without compromising the competitiveness of the organic sector. Obtaining, identifying, testing and physicochemical characterization of biologically active compounds with action to combat microbial diseases in vegetable crops shows both originality and complexity of activities proposed in the research work. The present study provides an overview of the current state of external inputs use and proposes a solution to the stricter European standards, by a systemic approach of biotechnological sciences and agricultural sciences, with immediate applicability of the obtained results in farm practices.

Key words: organic systems, vegetables, natural products, crop protection, plant products.

INTRODUCTION

The continuous growing of consumers demand for organic food lead to the need to develop solutions for increasing both production and quality obtained from organic systems.

Organic farming is a cultivation method that does not use pesticides or chemical fertilizers, and that replaces them with other methods including products obtained from plants or animals: purine, compost, mulch.

Organic production is defined as a holistic system that integrates cultural, biological, and mechanical practices that foster cycling of natural resources adapted to local conditions, promote ecological balance by avoiding external inputs with high resilience, and conserve biodiversity (IFOAM, 2009, Regulations of department of Agriculture US, 2011).

Development of pest and disease resistance to conventional pesticides and environmental contamination problems created pressure on farmers to adopt new strategies for disease control in horticultural production, both for conventional and organic systems.

It is expected that products obtained from biologically active compounds extracted from plants will play an increasingly important role in future crop protection strategies. The exploitation of antimicrobial substances from plants that inhibit or halt the reproduction of pathogenic microorganisms, would become a more realistic and ecological method for the integrated management of plant diseases and pests with the final goal of reducing or gradually phasing out contentious inputs, without compromising the competitiveness of the organic sector.

The present study provides an overview of the current state of external inputs use and proposes a solution to the stricter European standards, by a systemic approach of biotechnological sciences and agricultural sciences, with immediate applicability of the obtained results in farm practices. The high degree of originality and innovation in organic research lies in testing by detailed analysis of the designed solutions and its conceptual framing in standards provided by Directive 91/414/CEE, 15 July 1991 for Plant Protection Product. This study puts together, in an original way, both material resources from the field of vegetable crops and from technology aimed to decrease the microbial diseases attack.

SUSTAINABLE STRATEGIES USED IN PLANT PROTECTION MANAGEMENT

Biological control

Biological control is defined as the use of living organisms to control pests or the use of microbial antagonists to suppress diseases. There is a growing demand for biologically based pest management practices. Recent surveys of both conventional and organic growers indicate an interest in using biocontrol products suggesting that the market potential of biocontrol products will increase in the future (Heydari et al., 2010).

Biocontrol agents may be predatory, parasitic, or pathogenic; they may also be either "natural" (from naturally occurring organisms in the ecosystem such as wild beneficial insects) or "applied" (meaning external organisms and microorganisms that are introduced in the agrosystem). Biocontrol agents include insects, mites, bacteria, fungi, viruses, and nematodes.

In the recent years, extensive research is being carried on beneficial bacterial isolates that proved to have antimicrobial effects against various phytopathogens.

In vivo tests pointed out *Paecilomyces variotii* as a broad spectrum biocontrol agent effective against both bacterial spot of tomato and *Fusarium* wilt of melon (*F. oxysporum* f.sp. *melonis* and *X. campestris*) (Suarez et al., 2013).

Several fungal and bacterial biocontrol agents have been used as seed and soil application to reduce the incidence of plant diseases caused by soil borne fungal pathogens (Heydari et al., 2010).

Plant extracts

Traditionally, the plant extracts were used in the form of plant extracts and volatile oils. Aromatic and medicinal plants have attracted interests in the field of plant disease control, particularly plant extracts with antimicrobial properties and contain a spectrum of secondary metabolites such as alkaloids, quinolones, flavonoids, glycosides, saponins, tannins and terpenoids (Sales et al., 2016). Reports on medicinal plants extracts have shown *in vitro* inhibitory effects against phytopathogens (Rodino, 2013; Harlapur et al., 2014; Prashith K.T.R. et al., 2010). Plant extracts have also been reported to have antimicrobial properties against plant pathogens including viruses, fungi, bacteria, nematodes and insects *in vitro* as well as *in vivo* (Duraisamy, 2015).

Allelopathy

The term of allelopathy was first used in 1937 by Molisch, defining the chemical interaction (both stimulation and inhibition) between all types of plants, including microorganisms (Kalinova J., 2010; Bostan et al., 2013). Allelopathic plants are basically used as catch crops or trap crops. They are used in plant protection of tropical regions against parasite weeds, reducing the parasite seed bank by 72%. Allelopathic compounds can act as insects repellent (Kalinova J., 2010). Good results were also obtained for parasite weed control.

Intercropping

Intercropping is the planned mixed growing of more than two species at one area of land in the same period of time. A prerequisite for the success of intercropping is the interdependence of the selected crops in growth and development due to their biological particularities. The method is based on complex interactions between companion species with good results in trap cropping, weed suppression, physical-spatial interactions.

Intercropping effects are not only reducing pest populations, it proved to be efficient. in controlling plant diseases. The grain intercrop reduced humidity in the canopy and reduced the raindrop splash effect, those two conditions being the favorable to fungal spores spread (Schoeny et al., 2008). Moreover, studies showed that intercropping system is more productive than single cultivation system due to complementing effects of the companion crops.

CURENT CHALLENGES

The complexity of organic farming principles requires farmers to achieve a high level of knowledge and skills. The resources available in organic agriculture are presently limited due to very strict rules in force. Basically, crop protection solutions are limited to Integrated Pest Management solutions and the use of crop protection chemicals is drastically fenced, being allowed only few basic substances.

Basic substances allowed in organic agriculture are set by EU commission, and they consist of materials that are covered by the definition of food stuff Regulation (EC) No 178/2002 and have plant or animal origin. Basic substance is defined as active substance simultaneously fulfilling the criteria:

a) is not a substance of concern;

b) does not have an inherent capacity to cause endocrine disrupting, neurotoxic or immunotoxic effects;

c) is not predominantly used for plant protection purposes but is useful in plant protection either directly or in a product consisting of the substance and a simple diluent;

d) is not placed on the market as a plant protection product (EC, 2009).

Products based on natural components used against phytopathogens gained attention in search of environmentally friendly solutions usable either for mass production or for organic and low-input farming systems. After validation of antimicrobial effects of medicinal plants used in traditional medicine, the research aim expanded to evaluation of plant based products against phytopathogens (Rodino et al, 2013). Research on medicinal plants extracts and oils showed the inhibitory effects against various phytopathogenic fungi

In vitro studies on some species of plants such as weeds or trees with less known medicinal value, that could be used to control plant diseases and pests were also reported (Srivastava et al., 2011; Mahlo et al., 2010).

It was demonstrated the antifungal bio-capacity on powdery mildew, downy mildew and botrytis of grape cane extracts in greenhouse and field assays (Houille B. et al., 2015). These extracts were obtained after a long alcoholic extraction by hot soxhlet using large amounts of solvents. A recent study (Soural I. et al.,

2015) describes best extraction techniques, but the results show importance of methanol, a toxic laboratory solvent not usable in green technology and prohibited in the homologation category as basic substances.

FUTURE TRENDS

Scientific knowledge on the preparation mode of plant based products and how to use them are rather rare and incomplete. Therefore, it is important to optimize the extraction and manufacturing process, the stabilization of these preparations, dates and rates of application.The support of the use of these natural preparations or substances, as well as their market authorisation, requires their evaluation and registration within Plant protection Products Register. Plant protection product data must be collected from research on alternative or traditional crop protection methods (Marchand, 2015).

Another promising hypothesis for organic farming is the use of bacterial preparations based on nitrogen-fixing microorganisms. They are introduced into the root zone of plants, thus enriching the soil with biological nitrogen.

CONCLUSIONS

Development of pest and disease resistance to conventional pesticides and environmental contamination problems created pressure on farmers to adopt new strategies for disease control in horticultural production, both for conventional and organic systems.

Obtaining, identifying, testing and physico-chemical characterization of biologically active compounds with action to combat microbial diseases in vegetable crops implies both originality and complexity of activities proposed in the research work.

Moreover, plant protection product data must be collected from research on alternative or traditional crop protection methods.

Preliminary trials concerning antimicrobial potential of plant extracts are encouraging but full characterization and optimization is still required. However, to be fully efficient at farm level, proof of concept functionality has to be tested, and technical data sheets have to be produced.

ACKNOWLEDGEMENTS

This work was supported by a grant of the Romanian National Authority for Scientific Research and Innovation, CNCS/CCCDI - UEFISCDI, project number PN-III-P2-2.1-PED-2016-1544, contract number 14PED/2017, within PNCDI III.

REFERENCES

Bostan C., Butnariu M., Butu M., Ortan A., Butu A., Rodino S., Parvu C., 2013. Allelopathic effect of *Festuca rubra* on perennial grasses, Rom. Biotech. Letters, 18(2), 8190-8196.

Duraisamy S.,Loganathan K.,Rajendran R.,Kuppusami P.,Thiruvengadam R., 2015. Characterization of BioactiveCompounds from Botanicals for theManagement of Plant Diseases In Book Sustainable Crop Disease Management using Natural Products, Cabi.

EC (2009) Regulation (EC) No 1107/2009 of the European Parliament and of the Council of 21 October 2009, Official Journal of the European Union L309, 1–50.

Harlapur S.I., Kulkarni M.S., Wali M.C., Srikantkulkarni H., 2007. Evaluation of Plant Extracts, Bio-agents and Fungicides against Exserohilum turcicum (Pass.) Leonard and Suggs. Causing Turcicum Leaf Blight of Maize, India, J. Agr. Sci.20(3), 541-544.

Heydari A., Pessarakli M., 2010. A Review on Biological Control of Fungal Plant Pathogens Using Microbial Antagonists. Journal of Biological Sciences, 10: 273-290.

Houille B., et al., 2015, Biosynthetic origin of E-resveratrol accumulation in grape canes during postharvest storage. Journal of Agricultural and Food Chemistry 63, 1631-1638.

International Federation of Organic Agriculture Movements (IFOAM 2009).

Kalinova J., 2010, Allelopathy and Organic Farming, Sociology, Organic Farming, Climate Change and Soil Science Sustainable Agriculture Reviews, 3, 379-418.

Mahlo S.M., McGaw L.J., Eloff J.N., 2010. Antifungal activity of leaf extracts from South African trees against plant pathogens, Crop Protection 29, 1529-1533.

Marchand P.A, 2015. Basic substances: an opportunity for approval of low-concern substances under EU pesticide regulation Pest Management Science, 71 (9), 1197–1200

Prashith K.T.R. et al., 2010. Screening of selected single and polyherbal Ayurvedic medicines for Antibacterial and Antifungal activity, Anc Sci Life, 29(3),22–25.

Regulations of the Department of Agriculture, 2011. Agricultural Marketing Service (Standards, Inspections, Marketing Practices), Organic Foods Production Act Provisions, 7 CFR 205.2 - Terms defined.

Rodino S, et al., 2013. Investigation of the antimicrobial activity of extracts from indigenous *X. strumarium* plants against *P.infestans*, Curr Opin Biotech, 24, S72-S73.

Sales, M.D.C.; Costa, H.B.; Fernandes, P.M.B.; Ventura, J.A.; Meira, D.D., 2016. Antifungal activity of plant extract with potential to control plant pathogens in pineapple. Asian Pacific J. Trop. Biomed., 6, 26-31.

Schoeny A., Menat J., Darsonval A., Rouault F., Jumel S., Tivoli B., 2008. Effect of pea canopy architecture on splash ispersal of *Mycosphaerella pinodes* conidia. Plant Pathology. Vol. 57., 1073–1085.

Soural I., et al., 2015. Various Extraction Methods for Obtaining Stilbenes from Grape Cane of *Vitis vinifera* L. Molecules, 20, 6093-6112.

Srivastava D., Singh P., 2011. Antifungal Potential of Two Common Weeds against Plant Pathogenic Fungi, Asian J. of Biological Sciences, Volume 2(3), 525-528.

Suárez-Estrella F., Arcos-Nievas M.A.,. López M.J, Vargas-García M.C., Moreno J., 2013. Biological control of plant pathogens by microorganisms isolated from agro-industrial composts, Biological Control, 67, 3, 509–515

16

POSTHARVEST HANDLING OF FRUIT AND VEGETABLES IN TURKEY

Kenan KAYNAŞ[1]

[1]Çanakkale Onsekiz Mart University. Faculty of Agriculture. Terzioglu Campus. 17020. Çanakkale. Turkey

Corresponding author email: kenankaynas@gmail.com

Abstract

Turkey is the unique country that has land in both Europe and Asia. Turkey has approximately 75 000 000 population and is an agricultural country. In this context Turkey has great fruit and vegetable production potential. Turkey is one of the biggest producers of apples, pears, quinces, peaches and apricots, citrus (as fruit species) and tomatoes, peppers, melons, watermelons (as vegetable species). However, the amount of cold stores and packinghouses is deficient for its potential. Moreover the export rate is 5-10% of production. In addition the rate of postharvest losses of these crops is approximately 35-40%. On the other hand, the usage of controlled atmosphere (CA) stores and new postharvest technologies debouched rapidly. As a result increasing the amount of CA stores and modern packinghouses, arranging the errors in cold chain and improving the reproducing conditions will make Turkey an important fresh perishable exporter. Besides, the potential and capacity of production will be validated by minimizing the postharvest losses. On the other hand, cave storage will be improved by some modifications such as temperature, humidity and air circulation conditions. Their numbers, and hence capacities are steadily increasing. Eventually, the cave stores have impressions as being the predominating stores for citrus, potatoes and onions.

Key words: Turkey, horticultural production, postharvest losses, postharvest technology.

INTRODUCTION

STATUS OF HORTICULTURAL SECTOR

Turkey is blessed in the production of wide range of fruits and vegetables of which many are indigenous to the area such as pears, quinces, cherries, plums, grapes, walnuts, hazelnuts and pistachios. Of the total cultivated land, fruits and olives trees including vineyards occupy 3.012 million ha (11.0%) whereas vegetables 0.663 million ha (2.4%). It is estimated that 75% of vegetable fields and probably 40-50% of fruit, vine, olive orchards are irrigated. Although the land has nearly remained stable over the last decade the total production of horticultural produce increased from approximately 24 million tons to over 45 million tons (1990-2010). This may have been largely due to increase in large bearing orchards, and for vegetables, improved cultural techniques and high yielding varieties. Table 1 and Table 2 show figures on main fresh fruits and vegetables. Total annual fruit production was 9 million tons in 1990 and is probably around the 15-16 million tons in recent years. Total annual vegetable production was around 22 million tons in 1990 and is probably 30

million tons in recent years. Estimated export rate of 5-7% for fruits and vegetables reveals 1.5 million tons fruits and 1.8 million tons of vegetables are exported. Intensive production of fruits and vegetables is concentrated in the Mediterranean, Aegean, and Marmara Region as well as the central plateau, hazelnuts in the Black Sea Region, and pistachio in the Mediterranean and Southeastern Anatolia Region. Protected cropping is concentrated in microclimatic zones along the Mediterranean coast and Aegean Region. In these areas 90% of the total protected culture is used for vegetables, 7% for cut flowers and indoor plants, and 3% for fruits yield (strawberries, bananas, peaches).

The objectives of the Government for the development of horticultural sector have been mainly concentrated on: a) Modernizing production techniques to increase productivity and growers income, b) Producing highest quality of fruits and vegetables as possible, c) Meeting the food requirements of persistently increased population, d) Promoting horticultural exports as a result of oversupply.

In fact, many economic policies and reforms initiated by the Turkish government in the early

1990's and reviewed in 2002 have encouraged growers, private sector and market forces as well as exporters, as a result of which overall horticultural production has reached 45 million tons a year.

Table 1. Fruit production in Turkey (tons).

Year	Apple	Pears	Quince	Louquats	Peaches	Plums	Apricot	Cherries	Sour cherries	Cornel	Olive
1990	1900000	413000	79000	9000	350000	188000	240000	143000	90000	17000	1100000
2000	2400000	380000	105000	11500	430000	195000	530000	230000	106000	12000	1200000
2010	2600000	380003	121085	12112	539403	240806	450000	417905	194989	12517	1352827

Year	Almonds	Hazelnuts	Walnuts	Chestnuts	Pistachios	Mulberry	Pomegranates	Strawberries	Persimmon
1990	46000	375000	115000	80000	14000	80000	50000	51000	10000
2000	47000	470000	116000	50000	75000	60000	59000	130000	12000
2010	55938	600000	178142	59171	128000	75096	208502	299940	26277

Year	Oranges	Mandarin	Lemons	Grape fruits	Kiwi fruits	Bananas	Figs	Grapes	GrenTea Leaves
1990	735000	345000	357000	33000	-	36000	300000	3500000	608440
2000	1070000	560000	460000	130000	1400	64000	240000	3600000	758038
2010	1710500	858699	787063	213768	26554	210178	254838	4255000	1305566

Table 2. Vegetable production in Turkey (tons).

Year	Tomatoes	Cucumber	Pepper	Eggplant	Okra	Squash	Pumpkin	Melon	Water melon	Pea (green)
1990	6000000	1000000	900000	735000	22000	294000	57000	1650000	3300000	37000
2000	8890000	1825000	1480000	924000	27500	260000	72000	1865000	3940000	48000
2010	10052000	1739191	1986700	846998	36748	314340	89368	1611695	3683103	90191

Year	Bean (green)	Calavence	Cowpea	Broad beans	Cabbage	Lettuce	Artichokes	Celery	Cauliflower	Broccoli	Spinach
1990	430000	31000	-	62000	699000	186000	10000	12000	68000	-	160000
2000	514000	41000	12000	45000	725000	333000	24500	16500	90000	-	205000
2010	587967	70614	16591	41929	693012	419298	29070	1534	158579	26493	218291

Year	Cultivated mushroom	Onion	Garlic	Leek	Carrots	Radish	Purslane	Parsley	Mint	Dill	Rocket	Cress
1990	-	1736000	95000	340000	168000	71000	4000	-	-	-	-	-
2000	-	2428000	102000	308000	235000	167500	2250	40000	5000	1700	1150	1250
2010	21559	2065478	98170	244812	533253	155673	4936	56332	11772	2978	4058	2380

These measures and programs have an "open-end" potential (increases 100%, 200% or more) in contrast, however to "close-end" potential defined as postharvest technology and handling.

After realizing that estimated losses of horticultural crops running at 25-30% in Turkey more emphasis has been given to the postharvest handling systems in the last two decades.

POSTHARVEST SYSTEMS AND HANDLING PRACTICES

Postharvest systems and handling was first introduced in Turkey in the early 1960's by Prof. L. Lary Claypool from the University of California Davis. Now many people working at the Universities and Ministries have specialized in the postharvest field and carrying out research and training programs on horticultural crops.

In Turkey, most fruits and vegetables the higher temperature during the growing period the earlier the time of harvest. Long hot summers especially predominating in the South create sunburn problems on produce, thus detracting from quality. Sunburned crops may show symptoms of sun scald if they are stored too long like apples, pears, tomatoes, peppers etc. As a result some pome fruit varieties on dwarf rootstocks are growing with shading net systems recently. Russeting is another environmental problem in coastal regions. If leaf overlapping cannot avoided, tomatoes and many fruits as well as citrus, pome and stone fruits are highly affected for quality. Frost and hail damage cripple the yield and quality on many fruits and vegetables in some years.

Overhead sprinkling and mixing overhead fans are used for late spring cold injury especially.

More recently, Ministerial and some private organizations have development laboratories and mobile analyses units to help growers solve nutritional problems. But some postharvest physiological disorders connected with nutritional imbalances have a big problem still especially for pome fruits storage. All the calcium deficiency related disorders which appear during storage and during marketing have nearly been prevented. Foliar application of their salts has been a common practice by the growers on apples, pears, quince, tomatoes and pepper.

At the moment, extensive research studies conducted by the universities as well as by the Ministries Institutes have well established the use of growth regulators to increase produce quality. But, the use of growth regulators may have limited use due to harmful effects on human health, in practice. Also there are well established toxicology laboratory in some regions, for residue analyses on exporting fresh crops.

Many research studies are in progress relating preharvest cultural practices to the postharvest produce quality. Variety evaluations, dwarfing rootstocks, effects of pruning and thinning, mulching, soil cultivation and pollination are some areas being explored and the results are disseminated to growers and extension personnel for useful implementations.

Today, in Turkey, as in the past this is achieved through the hand harvesting in all horticultural crops except in the processing industry. There is no doubt that humans can accurately select for maturity, allowing accurate grading and multiple harvest, can handle the commodity with a minimum damage. Mechanical harvesting is practiced to great deal on sweet corn, potatoes, onions, olive, tomato paste production, juice and canning production.

Postharvest losses occurring quantitatively and qualitatively in all phases of post-production cycle (cold chain) are in the vicinity of 35-40%. Fruits and vegetables are generally only cooled by packers and exporters when products are to be transported over long distances. Cooling operations are nevertheless simple and carried out mostly in cold – rooms and far from being technologically development. Since much of the produce is locally sold, cooling is not commonly practiced, because it is expensive and losses are not considered excessive. The cold chain from producer to consumer is frequently interrupted due to lock of efficient facilities such as packing houses, cold stores, cold transport and cool market operations and distributions. Following the harvest fruits and vegetables in Turkey are destined for either storage or market. Preparations for both destinations are mostly done in the field, in the packing shed or in a covered area. Preparations include, receiving, cleaning, trimming, sorting, hand grading, sizing and packing (Figure 1).

Figure 1. Classical fruit cold storage, handling and packaging for apple

Only small production goes through the modern packinghouse operations where the sequence of operations varies with different crops packing houses were mostly in operation nearby big consumption center and also founded nearby shipping port. Major provinces were İzmir, Antalya, Mersin and Istanbul which nearly equally shared the total by 40% each. In general, sorting and sizing of products are made according to their physical properties such as diameter and weight. Some of them have been graded objectively the produce by color. So, many modern packinghouses mostly handling vegetables are operative in Marmara Region. They are equipped with modern machinery and annexed with cold storage. They are mostly owned and operated by the exporting companies. Many cold stores and packinghouse build to handle 5 000 – 10 000 tones in during the last 10 years, especially to handle postproduction phase of pome and stone fruits.

Some forms of deterioration, such as sprouting, water loss, storage disorders, insect manifestations and fungal rots can be minimized with chemical treatments before storage or marketing. Antitranspirants, surfactants and other skin coating agents and ethylene absorbents such as 1-methylcyclopropene (1-MCP), aminoethoxy vinyl glycine (AVG) are being investigate by research laboratories and their commercial use is to be spread. Such studies implicated plant nutrition studies and research efforts have yielded results of practical implementations. Field applications are widely practiced but their postharvest use before storage is limited due to lack packinghouse operations or simple machinery which can be used in packing sheds after harvest. Along this line, commercially scald is significantly reduced by the postharvest dip in diphenylamine (DPA) especially Granny Smith apples and domestic pear cultivar Deveci, Anjou and Abbe Fetel. Nevertheless, the growers in Turkey are showing keen interest on pre-storage treatments with chemicals including the fungicides, since fungal rots contribute to postharvest losses at the highest rates.

Hydrocooling and forced-air cooling systems are used by several modern cold stores in Turkey (Figure 2). Turkey could gain great

advantage their upon reduce postharvest losses especially during marketing if wholesale and regional markets as well as ever increasing supermarket chains are furnished with compact ice-bank cooling units.

Figure 2. Hydrocooling for cherry and DCA (Dynamic Controlled Atmosphere) atmosphere storage for apple

Curing of potatoes, onions and garlics is done in the field in Turkey where harvest time is characterized by hot and sunny days with concurrent low humidity levels. It is reasonable to believe that such in-field curing may create problems resulted from excessive heat, lack of aeration, soil-born diseases and field rodents and other pests. These factors obviously shorten the storage life and contribute the high rates of postharvest losses likely to occur in these crops at the vicinity of 35-40%. Proper temperature, humidity and air rate levels required for curing of onion and garlic have been obtained but these are limited practically.

The purpose of fresh fruits and vegetables storage in Turkey is not different than anywhere else in the world. Storage of fresh fruits and vegetables prolongs their consumption and in some cases even improves their quality. Even using the most modern cold stores, expectations of growers or handlers in Turkey sometimes are crippled due to high rates of storage losses because they ignore the fact that each member ring of the cold chain should be tightly bound to the next one and storage alone, as a separate ring, cannot improve the condition unless other former rings i.e., preharvest factors, harvesting and handling practices, precooling, packaging and hauling have been orderly and properly fulfilled.

Mechanically refrigerated stores capacity is about 750.000 tons in Turkey. Controlled atmosphere stores are a new concept for Turkey but it is progress because of storage quality, prolongs their consumption and price of crops at off-seasons. Only apples are stored in CA for 9 months (Figure 2).

The simplest of the alternative storage source but least practiced in Turkey is keeping the produce on the plant or, in situ for few months after they attain maturity. This is particularly applied on few citrus species, persimmon, grapes in certain areas, and among the vegetables, potato, carrot and garlic. Grapevines are individually covered with clear plastic as protection from early frosts.

Usually, some vegetables are stored in pits and trenchers by covering their surface with soil. Pits are used for storing potatoes, carrots, turnips and lesser extend on cabbage.

In some areas insulation with straw, hay or even manure is practiced when cold winter prevails. These varieties are mostly local and have no commercial value.

Cave storage can play an important role in storing some durable fruits and vegetables in regionally "Cappadocia Valley" which is historical place (Figure 3).

Their numbers, and hence capacities are steadily increasing.

Figure 3. Cave stores in Cappadocia Valley for potatoes storage

Eventually, the cave stores have impressions as being the predominating stores for citrus especially for lemons.

Much of the produce of lemons, oranges and grape fruit grown in the Mediterranean Region is shipped out to the area for subsequent storage. It is customary the hose the surfaces inside the store with water to provide extra humidity apples are to be stored. The floor is also watered occasionally. Potatoes and onions are kept in rather less humidified caves in different dimensions. Walls are usually 1.0-1.5 meters wide. Aeration is achieved by natural convection. The doors are opened early in the morning when the ambient air is cold and entering air flowing around the stacks of produce, leaves the cave through the pipe flues. Temperature inside the cave should not deviate so much and probably stays near the average annual ambient temperature. The overall storage capacity of these caves is near 700.000

tons all located within the provincial borders of Cappadocia Region To reduce overall postharvest losses in these cave stores some modifications appear to be essential. A refrigeration, heating unit and mechanically ventilation and inside circulation of air are indispensable.

The present situation of the postharvest system in Turkey is as follows: a) Small scale production mostly on fragmented land involves high physical handling, transportation and transaction costs, b) Postharvest losses are as high as 30%, c) Too many intermediate agencies in marketing the produce each demanding payment for their services, handling small lots generate overall high cost, d) Unstable market demands and prices create significant risks for market oriented produce.

CONCLUSIONS

Recommendations for future strategies should be as follows: a) The Government at first sight should make an extensive international market survey, implicating the foreign marketing research agencies. European Community (EC) countries, Middle East and Near East, Russia and Turkic States could constitute potential markets, b) Growers will orient themselves to produce high quality crops whose postharvest handling will demand extreme care and involve modern concepts of postharvest treatment, processing storage and transportation, c) In Turkey usually apples, pears and table grapes

are stored in cold stores. It should be cold stored in commonly different fruits and vegetables such as quince, peach, melon, cabbage, tomatoes, pepper, onion and potato etc. And store owners to modify their old stores into CA systems.

REFERENCES

Anonymous, 1993. Turkey –Horticulture Subsector Review. FAO Investment Center, FAO / World Bank Cooperative Programme. Ministry of Agriculture and Rural Affairs. Ankara.

Anonymous, 2010. Agriculture Structure and Production. State Institute of Statistics. Prime Ministry, Republic of Turkey. Pub.No.3455, Ankara.

Gündüz M., 1993. Importance of Cold Chain on Fresh Fruit and Vegetable and Studies on its Present State. Export Promotions Center (IGEME), Ankara., (in Turkish).

Kader A.A., 2009. Handling of Horticultural Perishables in Developing vs. Developed Countries. Proceedings of the Sixth Int. Sym., Acta Hort.877, V.1, 121-127.

Özelkök, S., 1987. Cool Transportation of Fruits and Vegetables. Refrigeration Technology in Food Storage and Processing Seminar, Istanbul Chamber of Trade (ITO). Pub. No. 1988-33, Istanbul, (in Turkish).

Özelkök S., Kaynaş, K., 1991. Postharvest Losses of Fruits and Vegetables and Their Remedies, Journal of Min. Agric. (TOK) 59,9-12.

Türk R., 1993. Horticulture Subsector Review. Study 1. Markets and Marketing. Postharvest and Transport of Horticultural Products. FAO Investment Center, FAO / World Bank Cooperative Programme. Ministry of Agriculture and Rural Affairs. Ankara.,

17

TECHNICAL ASPECTS CONCERNING THE PRESERVATION OF PEPPERS IN DIFFERENT STORAGE CONDITIONS

Marian VINTILĂ, Florin Adrian NICULESCU

Research and Development Institute for Processing and Marketing of the Horticultural Products - Bucharest, No. 1A, Intrarea Binelui Street, District 4, 042159, Bucharest, Romania,
E-mail: horting@gmail.com

Corresponding author email: marian.vintila57@yahoo.com

Abstract

The research concerns the study of the ability to maintain nutritional and commercial quality to peppers in different temperature storage conditions. During the three years of experimentation were used varieties: 'Buzau 10', 'Galben superior' and 'Bianca'. These were stored after proper preparation, at ambient temperature (+20 ... +22°C), refrigerated (+10 ... +12°C) and cold (+3 ... +5°C). The duration of storage, the weight (mass) and losses degradation and evolution of chemical components caused by the 9 variants yielded conclusive conclusions to 'Galben superior' and 'Bianca' cultivars for which experimental three-year cycle. It was found that the losses observed during storage are influenced by storage temperature and climatic conditions for the development of cultivars that have varied significantly in the three years of experimentation. The experimental variants were considered existing conditions in the family farms

Key words: peppers, quality preservation, storage, family farms.

INTRODUCTION

Maintaining the quality of horticultural products after harvest, pick still many problems especially for family farms, where the technical knowledge and the material is poor. Research has allowed the determination of the technical aspects of storage cultivar pepper in different storage conditions may be applied in family farms.
Pepper is perishable product storage. Pepper cultivars are grown in almost all households and vegetable farms for their own consumption and / or for sale. The two main factors responsible for maintaining the quality of peppers are: varieties resistance to storage and thermal conditions in storage areas. A series of researchers (Linda J. Haris 1998; Cantwell, I. M. and R.F. Kasmire 2002; Jamba, A. and B. Carabulea 2004; Thompson, F. J. and Crisosto H. C.2002) have studied the behavior of pepper in different storage conditions
Recent research conducted by "Horting" institute has watched highlighting storage resistance of some varieties grown in our country and the influence of temperature on the quality and duration of maintaining quality of peppers. The results may be indicative benchmarks for family farms with peppers in vegetable assortment.

MATERIALS AND METHODS

The study was taken in four varieties of peppers ('Buzau 10', 'Galben superior' and 'Bianca') grown in the same farm and placed in storage. Storing was carried out in three different conditions: ambient temperature (+20...+22°C), refrigerated spaces (+10...+12°C) and cold conditions (+3...+5°C). It covers such major environmental conditions in which the products in question may be kept in the household. Temporary storage after harvesting in different areas is carried out at ambient temperature, keeping the average in refrigerators or refrigerated rooms and long-lasting in cold storage facilities. It were determined the duration of preservation and level of weight (mass) and decay losses and the evolution of some chemical components during storage. The scheme of research organization that included nine experimental variants based on onion varieties and storage conditions is presented in table 1.

Table 1. The organization scheme
of research with peppers

Variant	Variety	Storage conditions
V1	Buzau 10	Ambient temp. (+20…+22°C)
V2	- idem -	Refrigeration (+10…+12°C)
V3	- idem -	Cold conditions (+3…+5°C)
V4	Galben superior	Ambient temp. (+20…+22°C)
V5	- idem -	Refrigeration (+10…+12°C)
V6	- idem -	Cold conditions (+3…+5°C)
V7	Bianca	Ambient temp. (+20…+22°C)
V8	- idem -	Refrigeration (+10…+12°C)
V9	- idem -	Cold conditions (+3…+5°C)

The main biometric data of peppers are presented in Table 2 and the appearance of chosen varieties in Figure 1.

Table 2. Biometric data

Variety	Length (height) (mm.)	Width (diameter) (mm.)	Shape index	Average mass (g/pcs)
Galben superior	83,8	64.6	1,30	90,99
Buzau 10	69,10	62,9	1,10	74,66
Bianca	77,3	58,8	1,31	89,16

Figure 1. Peppers appearance at the starting of experiments

The preparing stage of peppers for research purpose is illustrated in Figure 2.

Figure 2. Experience with peppers under preparation

RESULTS AND DISCUSSIONS

The average weight of the fruit is a characteristic indicator for every variety. The level of weight and decay losses and sprouting during storage in different temperature conditions are presented in Table 3.

Table 3. Losses accumulated during storage period (%)

Variety and other	Ambient Losses (%)			Refrigeration Losses (%)			Cold conditions Losses (%)		
	weight	decay	total	weight	decay	total	weight	decay	total
Galben superior	18.05	0	18.05	8.28	0	8.28	11.84	0	11.84
Buzau 10	8.91	0	8.91	6.26	0	6.26	6.69	0	6.69
Bianca	13.65	0	13.65	7.29	0	7.29	10.48	0	10.48
Mean	13.54	0	13.54	7.28	0	7.28	9.67	0	9.67
Storage time	5 days			10 days			20 days		

The data presented in Table 3 shows that in conditions of ambient temperatures, peppers can be kept up to 5 – 20 days (depending on variety), with average total losses of 13.60%. Buzau10 variety had the lowest total losses and 'Galben superior' variety showed the highest values of total losses. 'Buzau 10' variety proved to have better resistance to storage mainly because of reduced decay losses. Appearance of 'Buzau 10' variety of storage at ambient temperature is shown in Figure 3.

Figure 3. 'Buzau 10' variety after storage under ambient conditions

In refrigerated spaces peppers was stored for 10 days with 9.85% average total losses. In such conditions 'Buzau 10', was the most resistant cultivar with 8.28% total losses. On second place was situated 'Galben superior' variety, with total losses below the average of the three varieties studied. Last place was occupied by 'Bianca' variety, which cumulated the highest weight and decay losses. Appearance of peppers variety of storage at refrigerated temperature is shown in Figure 4.

Figure 4. Peppers variety after storage
under refrigerated conditions

Table 4. Level and evolution of chemical components in
peppers variety

Var.	Variety	Storage temp. (°C)	Soluble solids (%)	Acidity (%)	Total sugar (%)	Vitamin C (mg/100g)
	Galben superior	initially	6,5	0,15	2,22	121,32
V1		20-22°	6,1	0,14	2,18	36,70
V2		10-12°	5,1	0,13	1,89	19,57
V3		3-5°	4,5	0,10	1,72	16,22
	Buzau 10	initially	8,0	0,22	3,18	219,32
V4		20-22°	7,2	0,22	2,85	144,57
V5		10-12°	6,5	0,12	2,45	140,74
V6		3-5°	6,2	0,23	2,38	157,20
V7	Bianca	initially	6,5	0,19	2,79	168,36
		20-22°	6,0	0,14	2,40	65,20
V8		10-12°	5,5	0,18	2,44	104,32
V9		3-5°	4,7	0,12	2,09	122,97

In cold conditions, the average total losses raised to 10.61% after 20 days of storage. Thus the variety 'Buzau 10' were recorded the lowest weight anddecay losses cumulating 6.69% total losses, the variety 'Bianca' had about 10.61% and the variety 'Galben superior' had about 11.84%. Appearance of peppers variety of storage at cold temperature is shown in Figure 5.

Figure 5. Peppers variety after storage under cold
conditions

Content and evolution of some chemical components during storage are presented in Table 4. The data presented on Table 4 shows that initially the peppers had 6.5 to 8.0% soluble solids content, from 0.10 to 0.23% treatable acidity, from 1.72 to 3.18% total sugar and from 121.32 to 219.32 mg/100g vitamin C, depending on the variety. 'Buzau 10' variety had the highest content of soluble solids, treatable acidity, total sugar and vitamin C, while those of 'Galben superior' variety were recorded the lowest values of all components.

Main chemical components evolution was different from an experimental variant to another. The content of soluble solids had a downward trend for all the varieties in particular to the peppers stored in cold conditions. Values lower than initial ones have been recorded also by 'Buzau 10' variety (V5) and 'Bianca' variety (V8) stored under refrigeration conditions. 'Buzau 10' variety, considered the healthiest (lowest decay losses) showed a decrease in soluble solids and total sugar content in refrigeration and cold storage conditions. For 'Buzau 10' variety the soluble solids content remained high in refrigerated and cold conditions (V5 and V6) and for 'Bianca' variety only in refrigerated conditions (V9).

Acidity of peppers presented both slight increases and decreases depending on the variety and on storage conditions. The peppers maintained in general the initial acidity content in ambient and cold conditions and presented mild reductions in refrigerated one, the lowest values being recorded by 'Galben superior' variety (V3) and 'Bianca' variety (V9).

On ambient conditions total sugar content was maintained at high values at 'Galben superior' variety (V1) and 'Buzau 10' (V4) and had a decreasing trend at 'Bianca' (V7). In refrigerated conditions all varieties of peppers were significant reductions of total sugar content. And in cold conditions peppers from 'Buzau 10' variety (V6) and 'Bianca' variety (V9) had a total sugar content lower than initially, while the 'Galben superior' variety (V3) maintained a high sugar content.

The amount of vitamin C in peppers strongly decreased during storage at all varieties, but bet on in different proportions depending on variety and storage conditions. The sharpest decrease occurred in the 'Galben superior' variety of the genus to which vitamin C content decreased by 62-90% depending on storage conditions. Lowest losses of vitamin C were found in peppers of the genus 'Buzau 10' they were 30-50%, depending on storage conditions. Variety 'Bianca' has dropped by 44-74% vitamin C content according to storage conditions. In ambient conditions 'Galben superior' variety had lowest losses of vitamin C, while refrigerated and cold varieties 'Buzau 10' and 'Bianca' showed the lowest losses of vitamin C.

CONCLUSIONS

Maximum storage life of onions was 5 days under ambient conditions (depending on variety), 10 days under refrigeration and 20 days in cold conditions, with average total losses of 13.54%, 7.28 % and 9.67% respectively.

The optimum time to maintain the quality of the sweet is 3 days in ambient conditions, 7 days in refrigerated spaces and 15 days in cold spaces.

Pepper varieties have different behavior in similar conditions of storage. While in variety 'Buzau 10' bet on all storage conditions obtained the best results with the lowest volume loss, 'Galben superior' variety showed great sensitivity higher, posting the biggest losses in both ambient conditions and in cold.

Evolution of the main chemical components (dry matter, acidity, total sugar and vitamin C) of peppers can be an important indicator of the ability to maintain quality of storing bet on different conditions.

The best results maintain the quality of the peppers variety were obtained 'Buzau 10', which proved the most resistant, the best storage conditions are ensured by freezing at 10-12 °C.

The peppers must be checked daily to notice in advance of any change in appearance or quality and to intervene immediately to remove the causes or effects already produced.

The results have confirmed that the cold storage conditions recommended given the lowest loss (mass and impaired), and that increases shelf life, even up to 20 days in the case of peppers have a perishable high.

REFERENCES

Cantwell, I. M. and R.F. Kasmire (2002) - Postharvest Handling Systems, Postharvest Technology of Horticultural Crops (Chapter 35), Publication 3311, University of California.

Jamba, A. and B. Carabulea (2004) - Tehnologia pastrarii si industrializarii produselor horticole, Editura Cartea Moldovei, Chisinau.

Haris Linda J. (1998) – Peppers Safe Methods to Store, Preserve and Enjoy, Publication 8004, University of California.

Thompson, F. J. and Crisosto H. C. (2002) - Handling at Destination Markets, Postharvest Technology of Horticultural Crops (Chapter 21), Publication 3311, University of California.

18

INFLUENCE OF DIFFERENT ORGANIC MULCHES ON SOIL TEMPERATURE DURING PEPPER (*CAPSICUM ANNUUM* L.) CULTIVATION

Milena YORDANOVA[1], Nina GERASIMOVA[2]

[1]University of Forestry, Faculty of Agronomy, 10 Kliment Ohridski Blvd, 1756, Sofia, Bulgaria, GSM: +359.887.698.775, Email: yordanova_m@yahoo.com

[2]Institute of Plant Physiology and Genetics, Bulgarian Academy of Sciences, Acad. G. Bonchev Street, Bldg. 21,1113, Sofia, Bulgaria, Email: gerasimova_n@abv.bg

Corresponding author email: yordanova_m@yahoo.com

Abstract

The aim of the paper was to present the influence of different types of organic mulch on soil temperature during the cultivation of pepper. The experimental work was carried out in 2012-2013in the experimental field on University of Forestry – Sofia, with pepper cv. 'Sofiiskakapia'. For the purpose of the study were used different available materialsas organic mulches, which were waste products from organic agriculture:spent mushroom compost (SMC),barley straw (BS), grass windrow (GW), weeds. Mulched plots were compared with two control variants – hoed control plots, (HC) and non-hoed control plots (NHC).The mulching materials were spread manually in a 5-6 cm thick layer, after strengthen the seedlings of pepper. The soil temperature was recorded in 7 days, at a depth of 0, 5, 10 and 15 cm, by calculating the average daily temperature, from mulching to harvesting of production. Mulching materials affect soil temperature. Least variation in soil temperature was recorded at a mulch of straw with average temperature 19-22°C during the August, when the air temperature was highest. With the greatest variation in soil temperatures of mulching plots were two variants: with mulch of weeds and with mulch of grass windrow where the green waste materials, used for mulching, started to decompose slowly.

Key words: soil temperature, barley straw mulch, spent mushroom compost, grass windrow mulch, pepper.

INTRODUCTION

The optimum temperature for cultivation of bell pepper is between 18 and 25 °C. Extremely high temperatures above 32–35°C, in combination with a low air humidity, cause falling off of the flowers and fruit sets and increase the percentage of non-standard deformed fruits. Root system develops better when soil temperature is between 18–22°C (Panayotov et all. 2006).

In the hot summer days high soil temperatures affect evaporation and soil moisture, and hence the growth and development of pepper (Van Donk et al., 2011). Temperature stress, which is obtained at high soil temperatures at uncovered soil (32 – 34°C) may be minimized by the use of the mulch. (Godawatte et al., 2011; Yordanova and Gerasimova, 2012). Mulching improves plant growth, increased yields and quality (Sharma and Sharma, 2003; Singh et al., 2007). The organic mulches which are recycled into the soil can reduce the cost of production and are useful for the environment (Roe et al., 1992).

One of the best materials for mulching is compost, especially for growing of intensive crops (Vogtman, 1990; Yordanova, 2008). Mulching with grass also has a positive effect on the quality and quantity of the yields in a number of crops (Dvořák et al., 2009; Sinkevičienė, et al., 2009). It degrades faster than other mulch materials, with 10 cm layer of grass windrow is more effective than the 5 cm layer (Jodaugienė et al., 2012). Grass windrow positively affect the activity of soil enzymes and biomass in the soil (Jodaugienė et al., 2010).

It has been found that the mulching with straw has a favorable effect on the growth of pepper (Roberts and Anderson, 1994; Mochiah et al., 2012). This is explained by the preserving and maintenance of the soil moisture, the maintenance of a moderate soil temperature and suppressing the growth of weeds. Use of green stalks of weeds as mulch material only ismentioned as a method applied in old gardening practices historically. This provoked the decision to explore the possibility of using green weed residues as mulch material and to determine the influence of different types of mulch on soil temperature and crop yields in the cultivation of pepper.

MATERIALS AND METHODS

The experiment was conducted in 2012-2013, in the experimental field of the University of Forestry – Sofia (42°7′ N, 23°43′E and 552 m altitude). The soil is fluvisol, slightly stony, slightly acidic. This area came under a continental climatic sub region, in a mountain climatic region.

The study was performed with bell pepper (*Capsicum annuum*), cv. 'Sofiiskakapia', with growing period lasted 116-120 days, with pre-produced seedlings. Planting in the open field wass made on 21-22 May in both experimental years. Each plot was of 1.20 m wide and 3 m long. It contained two parallel rows of plants at 60 cm distance between rows and plants within rows were separated by 20 cm.

The experiment was carried out by randomized complete block design with six treatments and three replications. The tested treatments were: bare soil, maintained weed-free by hoeing – control plot(BSCP); non-mulched and non-hoeing (weeded) control plot (NMCP); mulch from spent mushroom compost (SMCM); mulch from barley straw (BSM); mulch from grass windrow (GWM); mulch from green stalk residues from weeds (WRM).

As weed residues mulch we used widely distributed weeds: common amaranth (*Amaranthus retroflexus* L.), fat-hen (*Chenopodium album* L.), and cockspur (*Echinochloa crus-galli* L.). In 2012 we used them separately, to could check if they will take roots again. In 2013 we used them mixed together as they grew naturally.

The mulches were applied to the soil surface by hands at a thickness of 5-6 cm, after the seedlings of pepper were strengthen – on 11 – 12 of June in both experimental years.

All elements of agrotechnical activities (basic and pre-sowing cultivation, irrigation, etc.) were the same for all treatments. The plants were irrigated by sprinkler irrigation system.

The soil temperature was monitored at soil surface (0 cm depth) and at a depth of 5, 10 and 15 cm, once of week, three times a day, throughout the period from the beginning of July till the end of September. The soil temperature was measured with hand-held needle soil digital thermometer.

Means were separated by application of Duncan's Multiple Range Test at $p \leq 0.05$.

RESULTS AND DISCUSSIONS

The average mean air temperature during the growing period (June – September) of both experimental years (2012-2013) was within the borders of optimum temperature of pepper cultivation, with one exception – on September, 2013, when the average mean temperature was 16.7 °C.

In spite of the optimum mean air temperature, the maximum air temperature, which was measured during the growing period, was higher than optimum for growing pepper (Table 1.).

Monitoring of soil temperature started in the beginning of July and continued until the end of September. This covered the hot summer period and lasted until the end of harvest. The maximum air temperature had an effect on bare soil temperature.

The average soil temperatures recorded in the middle of the August in all treatments and in four depths are presented in Figure 1. The soil temperature was measured at four depths (0, 5, 10 and 15 cm). We compared them with average daily air temperatures, to present the fluctuation in soil temperatures in different depths.

Table 1 Average air temperature (maximum, minimum, mean) and amount of precipitation during the growing period of the pepper for both years of the experiment.

	Average air temperature (°C)						Amount of precipitation (mm)	
	2012			2013			2012	2013
	max	min	mean	max	min	mean		
May	20.6	9.7	15.2	24.0	11.2	17.6	131.4	31.9
June	28.2	14.8	21.5	24.8	13.0	18.9	8.9	113.6
July	32.1	17.5	24.8	26.6	14.2	20.4	41.2	61.2
August	30.9	15.2	23.1	29.4	15.9	22.6	45.0	13.0
September	26.4	12.2	19.3	23.2	10.2	16.7	48.2	21.0

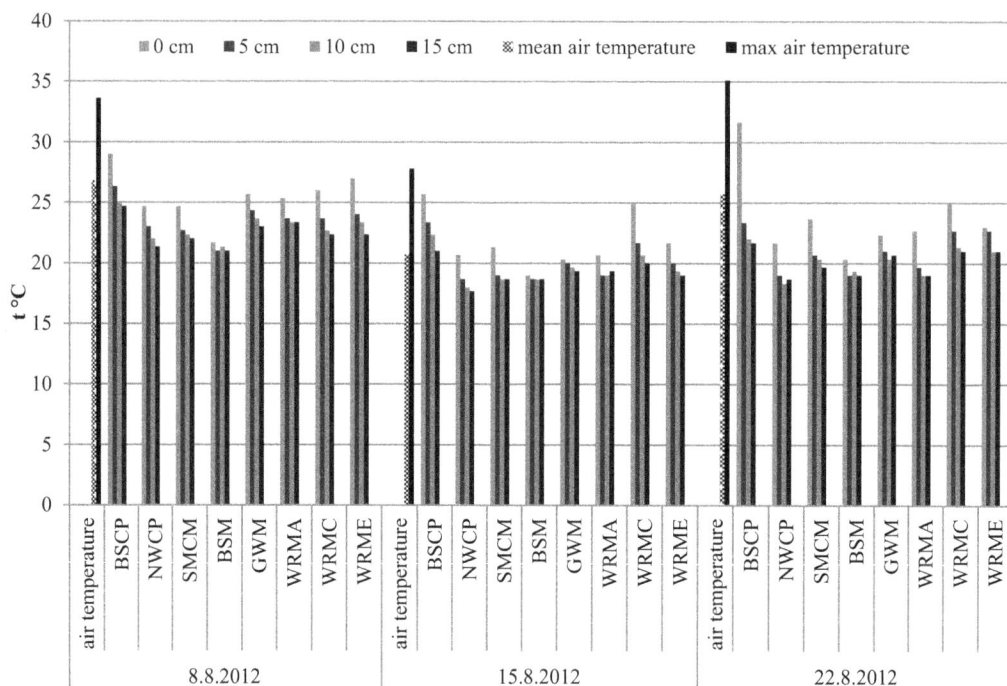

Figure 1 Average daily air temperatures, compared with average soil temperatures recorded in the middle of the August in all treatments and in four depths

The largest temperature differences are recorded at a depth of 0 and 15 cm. The temperature, recorded on the soil surface level of the bare soil control plot (BSCP) was always higher than temperature on the soil surface of the mulched plots (Table 2 and Table 3). However, temperatures of the soil surface of this option was not higher than the average air temperature, which is due to shading of the soil surface by the leaves of pepper.

In BSCP treatment were recorded and the highest temperature amplitudes between average daily soil temperatures on soil surface and at a 15 cm depth, which were significantly higher. The highest temperature amplitude (7.5°C) was recorded in July, 2012 (Table 2), which was the hottest month during the experimental period, the average soil temperature, was also higher (26.5°C) than optimum soil temperature for pepper roots (18-22°C).

The soil surface temperatures, measured at the mulched plots where always lower than the temperatures measured at the bare soil control plot. The soil temperature under different mulches was also affected by the type of used mulch materials.

Lowest soil temperature measured at the soil surface was recorded under barley straw mulch (BSM). This may be due to the fact that mulch of straw has a bright surface and reflects the sun's rays, preventing overheating of the soil surface. On the other hand keeps the temperature in the lower soil layers and was considered the smallest temperature range between 0 and 15 cm depth in the soil. The fluctuation in soil temperature at different levels is between 0.6 °C (in August, 2012) and 1.4 °C (in August and September 2013). Badaruddin et al., 1999, indicates that the use of straw mulch in areas with high temperatures maintaining the temperature in the soil below the day and higher during the night, and thus protects plants from heat stress. At this mulched treatment was recorded and lower average soil temperature (17.2 °C) than optimum soil temperature for pepper roots (18-22 °C).

In the treatment with weed residues mulch temperatures measured on the soil surface and a depth of 15 cm were higher in the first months, compared to other mulch options. This is due to the fact that under the influence of high air temperatures and moisture in the soil, fresh green plant matter begins to decompose, intensifying the microbiological activity of the soil surface. This leads to the maintenance of higher soil temperatures, as compared to other ground cover materials.

In 2012 we used green stalk weed residues (WRM) as mulch separately, to check if they will take roots again and to study them as a mulch material. The mulch from *Amaranthus retroflexus* L. (WRMA), the mulch from *Chenopodium album* L. (WRMC) and the mulch from *Echinochloa crus-galli* L. (WRME) were decomposed almost at the same time. They didn't take roots again and covered soil surface good in the beginning. The differences in measured soil temperature among these treatments are not significant, with one exception – the temperature amplitude, recorded at WRMA (2.6 °C) in August, 2012 was smaller than at other two treatments. In 2013 we used them mixed together as they grew naturally (Table 2 and Table 3).

At the end of the vegetation of pepper in 2012, mulch from weed residues (WRM) was wilted, was thin and because we used the whole stems, they are no longer covered well the soil surface and because of this the temperature at a depth of 15 cm in September was low and almost close to temperature variations without mulch .

Grass windrow was applied as mulch (GWM) after one – two days, not as fresh as weed residue mulch. The averages mean soil temperatures, which were recorded during the experimental years, were close to the borders of optimum soil temperature levels (17.6-22.2 °C).

Mulch from spent mushroom compost (SMCM)also maintained a higher temperature than barley straw mulch (Table 2 and Table 3). The dark color of this mulch affected the soil temperature and in September the averages mean soil temperatures, for both experimental years, were in the borders of optimum soil temperature levels (18.1-18.2 °C).

Table 2 Average daily soil temperature (at 0 cm and at 15 cm depth), mean temperature and temperature amplitude between two levels ($T_{amp}=T_{0cm}-T_{15cm}$) during growing period of pepper, 2012

| | Average soil temperature (°C) | | | | | | | | | | | |
| | July | | | | August | | | | September | | | |
	0 cm	15 cm	mean	T amp	0 cm	15 cm	mean	T amp	0 cm	15 cm	mean	T amp
BSCP	30.2a	22.7a	26.5	7.5a	28.0a	21.4 ns	24.7	6.6a	23.5a	16.8ns	20.1	6.7a
NMCP	22.3b	19.5b	20.9	2.8b	21.8b	19.0 ns	20.4	2.8c	18.3b	16.0 ns	17.2	2.3b
SMCM	23.0b	21.1b	22.1	1.9c	22.0b	19.8 ns	20.9	2.2c	19.3b	17.0 ns	18.2	2.3b
BSM	20.7c	19.6b	20.2	1.1d	20.0c	19.4 ns	19.7	0.6e	18.0b	16.8 ns	17.4	1.2c
GWM	23.2b	21.2b	22.2	2.0c	22.6b	20.8 ns	21.7	1.8d	18.3b	17.0 ns	17.7	1.3c
WRMA	23.4b	20.9b	22.2	2.5b	22.4b	19.8 ns	21.1	2.6c	18.8b	16.5 ns	17.7	2.3b
WRMC	24.2b	21.1b	22.7	3.1b	23.8b	20.4 ns	22.1	3.4b	17.8b	15.8 ns	16.8	2.0b
WRME	23.9b	21.0b	22.5	2.9b	23.4b	20.2 ns	21.8	3.2b	18.3b	15.8 ns	17.1	2.5b

Values with the same letter within years are not significantly different (Duncan's Multiple Range Test at p ≤ 0.05)

Table 3 Average daily soil temperature (at 0 cm and at 15 cm depth), mean temperature and temperature amplitude between two levels ($T_{amp}=T_{0cm}-T_{15cm}$) during growing period of pepper, 2013

| | Average soil temperature (°C) | | | | | | | | | | | |
| | July | | | | August | | | | September | | | |
	0 cm	15 cm	mean	T amp	0 cm	15 cm	mean	T amp	0 cm	15 cm	mean	T amp
BSCP	25.8a	19.0a	22.4	6.8a	27.6a	20.9ns	24.3	6.7a	21.2a	14.6ns	17.9	6.6a
NMCP	19.2b	17.9b	18.6	2.3b	21.2b	18.5ns	19.9	2.7b	18.0b	15.8ns	16.9	2.2b
SMCM	19.8b	18.1b	19.0	1.7c	21.8b	19.6ns	20.7	2.2b	19.1b	17.0ns	18.1	2.1b
BSM	18.2b	16.9b	17.6	1.3c	19.7b	18.3ns	19.0	1.4c	17.8b	16.4ns	17.2	1.4c
GWM	18.7b	17.6b	18.2	1.1c	22.0b	20.4ns	21.2	1.6c	18.2b	16.9ns	17.6	1.3c
WRM	18.8b	17.2b	18.0	1.6c	22.8b	20.1ns	21.5	2.7b	18.7b	16.7ns	17.7	2.0b

Values with the same letter within years are not significantly different (Duncan's Multiple Range Test at p ≤ 0.05)

The mulch protects the soil from strong overheating during midday hours and cooling at night, which explains the small temperature differences between the measured surface temperatures of the soil in mulched treatments and soil temperatures at a depth of 15 cm. This is in agreement with other researchers: Chen et al., 2004, 2005, 2007 demonstrated a reduction in the maximum temperature of the soil, increase of the minimum temperature, and a reduction in amplitude between day and night temperatures. Pinamonti, 1998, also indicated that mulching reduces fluctuations in soil temperature.

CONCLUSIONS

The bare soil surface is more affected by the air temperature and this is more strongly expressed by temperature fluctuations in the depth of the soil.

Mulches affected the soil surface temperature and keep it moderate. The soil temperature under different mulches was also affected by the type of used mulch materials.

With the least variation in the soil temperature is the treatment with mulch of barley straw, which maintains a temperature in the range of 19-22 °C in August and 17-19 °C during the first two ten-day periods of September.

With the wide variation in soil temperatures of mulch are two treatments: mulch from green weed residues and mulch from grass windrow. It was established in July and August, when the green weed residues and grass windrow used for mulching, decomposed.

Mulch of barley straw maintains moderate soil temperature and protects the soil from rapid temperature fluctuations in depth.

REFERENCES

Badaruddin, M., M. P. Reynolds and O. A. A. Ageeb, 1999. Wheat management in warm environments: Effect of organic and inorganic fertilizers, irrigation frequency and mulching. Agronomy Journal, 91:975-983.

Chen S.Y., X.Y. Zhang, D. Pei, H.Y. Sun, 2004. Soil evaporation and soil temperature in maize field mulched with wheat straw. J. of Irrigation and Drainage, 4.

Chen S.Y., X.Y. Zhang, D. Pei, H.Y. Sun, 2005. Effects of corn straw mulching soil temperature and soil evaporation of winter wheat field.Translocations of the Chinese Society of Agricultural Engineering, 10.

Chen S.Y., X.Y. Zhang, D. Pei, H.Y. Sun, S.L. Chen, 2007. Effects of straw mulching on soil temperature, evaporation and yield of winter wheat: field experiments on the North China Plain. Annals of Applied Biology, 150(3): 261-268.

Dvořák, P., K. Hamouz, P. Kuchová, J. Tomášek., 2009. Effect of grass mulch application on tubers size and yield of ware potatoes in organic farming. Bioacademy 2009 – Proceedings. Organic Farming – A Response to Economic and Environmental Challenges. Block I. Diversity in plant production, p. 35-37

Godawatte, V.N.A., C.S. De Silva and M.D.M. Gunawardhana, 2011.Effect of mulch on growth and yield of chilli (*Capsicum annuumL.*).pp. 8 http://digital.lib.ou.ac.lk/docs/handle/701300122/380

Jodaugienė, D., R. Pupalienė, A. Sinkevičienė, A. Marcinkevičienė, K. Žebrauskaitė, M. Baltaduonytė, R. Čepulienė, 2010.The influence of organic mulches on soil biological properties.Zemdirbyste-Agriculture, 2010, 97(2), p. 33-40

Jodaugienė, D., R. Pupalienė, A. Marcinkevičienė, A. Sinkevičienė, K. Bajorienė, 2012. Integrated evaluation of the effect of organic mulches and different mulch layer on agrocenosis.Acta Sci. Pol., HortorumCultus, 2012, 11(2), 71-81.

Mochiah, M. B., P.K. Baidoo, G. Acheampong, 2012. Effects of mulching materials on agronomic characteristics, pests of pepper (Capsicum annuum L.) and their natural enemies population. Agriculture and Biology Journal of North America. 3(6):253-261

Panayotov, N., S. Karov, R. Andreev, 2006. Organic production of pepper. Plovdiv, 68 p.(in bulgarian)

Pinamonti, F., 1998. Compost mulch effect on soil fertility, nutritional status and performance of grapevine. Nutrient Cycling in Agroecosystems, vol. 51, 3:239-248

Roberts, B. W. and J. A. Anderson, 1994. Canopy Shade and Soil Mulch Affect Yield and Solar Injury of Bell Pepper. HortScience 29(4):258-260

Roe, N.E., H.H.Bryan, P.J. Stoffella, T.W. Winsberg 1992. Use of Compost as Mulch on Bell Pepper. Proc. Fla. State Hort. Soc. 105: 336-338

Sharma, R.R. & Sharma, V.P. 2003. Mulch influences fruit growth, albinism and fruit quality in strawberry (Fragaria x ananassa Duch.). Fruits 58, 221–227.

Singh, R., S., Sharma, R.R. & Goyal, R.K. 2007. Interacting effects of planting time and mulching on "Chandeler" strawberry (Fragaria x ananassa Duch.). Sci. Hortic. 111, 344–351.

Sinkevičienė A., D. Jodaugienė, R. Pupalienė, M. Urbonienė, 2009. The influence of organic mulches on soil properties and crop yield.Agronomy Research 7(Special issue I), 2009, 485–491.

Van Donk, S. J., D. T. Lindgren, D. M. Schaaf, J. L. Petersen and D. D. Tarkalson, 2011. Wood chip mulch thickness effects on soil water, soil temperature, weed growth and landscape plant growth.Journal of Applied Horticulture, 13(2): 91-95

Vogtmann, H., 1990. Ecological gardening. The "Organic Agriculture", Plovdiv, page 96.

Yordanova, M., 2008. Biological studies of effects of using mulches in growing of broccoli - *Brassica oleracea* var. *italica* Plenck. PhD diss., University of Forestry, Sofia (in Bulgarian)

Yordanova, M., N. Gerasimova, 2012. Effect of mulching on soil temperature of onion (*Allium cepa* L.) Ecology and Health, Proceedings of the ninth national scientific conference with international participation.Academic Publishing Agricultural University. Plovdiv. 165-170

19

INFLUENCE OF GRAFTING ON PRODUCTION AT SOME GRAFTED EGGPLANTS

Mădălina DOLTU[1], Marian BOGOESCU[1], Dorin SORA[1], Vlad BUNEA[2]

[1]Research and Development Institute for Processing and Marketing of Horticultural Products – Horting, 1A Intrarea Binelui Street, District 4, 042159, Bucharest, Romania

[2]Central School, Bucharest, Romania, 3-5 Icoanei Street, District 2, 20451, Bucharest, Romania

Corresponding author email: doltu_mada@yahoo.com

Abstract

The scientific research is concerned continuously of achieving some high value biological creations, resistant or tolerant to diseases and pests, high productivity, high quality fruits; these goals are obtained by grafting. The eggplants are plants that can be grafted onto different rootstocks. The grafted seedlings transmit to crops quality, productivity, resistance to pests and diseases from soil, tolerance to abiotic stress factors, optimal absorption of water and nutrients, vigour. The study was conducted in a greenhouse of the Horting Institute Bucharest and it has followed the rootstock influence on the production quality at some grafted eggplants. The biological material has consisted of import F1 hybrids, eggplant scions ("Black Pearl" and "Classic") and rootstocks ("Emperador" and "Torpedo") commonly used in Romania for grafting of eggplants. It were determined form index (FI), form (F), weight/fruit (W), marketable production (MP), humidity (H) – gravimetric method, soluble dry matter (SDM) by refractometry method using ABBE refractometer, total sugar (TS) by Bertrand method. The results show that the grafting had influenced some aspects of the eggplant production. The conclusions of the researchers from the Horting Institute are in respect with some conclusions of the foreign researchers, but more researches are required in this domain to highlight the grafting effect on some aspects concerning production of grafted vegetables.

Key words: grafting, yield, Solanum melongena L., quality.

INTRODUCTION

The scientific research is concerned continuously of achieving some high value biological creations, resistant or tolerant to diseases and pests, high productivity, high quality fruits; these goals are obtained by grafting (Doltu, 2007).

The vegetable grafting has been used since the early decades of the XIX century in the countries from the Far East and it is considered to be an ecological way to reduce the attack of pathogens and pests of soil (fungi, bacteria and nematodes) which, particularly in intensive culture conditions, produce considerable production loss and abandonment of some cultures.

This planting material is a biological alternative for replacing polluting chemicals used to disinfection of the soil (Echevarria et al., 2004). The grafted seedlings print quality, productivity, resistance to diseases (*Fusarium* spp., *Verticillum* spp.) and pests (nematodes) of soil (Bogoescu et al., 2008).

The grafting print resistance to pathogens and pests of soil, tolerance to abiotic stress factors, improves absorption about water and nutrients, increase vigour to scion (King et al., 2010 and Lee, 1994).

Some researchers belive that the results concerning the fruit quality obtained from grafted plants are contradictory (Davis et al., 2008).

Çürük et al., 2009 had investigated the grafting influence on eggplants, noting that the fruit average weight is significantly influenced by grafting.

The results concerning the grafting influence on vegetable crops require more researches in this domain for to highlight the grafting effect on some aspects concerning production of grafted vegetables.

MATERIALS AND METHODS

The biological material used in research has consisted from eggplant scions and rootstocks commonly used in Romania to grafting of eggplants.

The scions were F1 hybrids of eggplants, 'Black Pearl' (Enza Zaden, Netherlands), 'Classic' (Clause Vegetale Seeds, France) and F1 hybrids of rootstocks 'Emperador' (Rijk Zwaan, US) and 'Torpedo' (Ramiro Arnedo Semillas, Spain).

The ungrafted and grafted eggplant seedlings had been obtained into specialized greenhouse for production of seedlings from Laboratory of Protected Cultures, Institute Horting Bucharest.

The grafting technique have consisted in more stages: sowing (scion and rootstock), preparing for grafting, grafting itself, introduction of grafted plants in polyethene tunnel for callus forming, transferring of seedlings in greenhouse for grower and maintenance according with the 'Classic' technology (Bogoescu M. et al., 2008).

The experimental variants were made up from lots of ungrafted (control) and grafted plants:

- ungrafted plants (control):
- 'Black Pearl' (V1)
- 'Classic' (V2)
  - grafted plants (scion x rootstock):
- 'Black Pearl' x 'Emperador' (V3)
- 'Classic' x 'Emperador' (V4)
- 'Black Pearl' x 'Torpedo' (V5)
- 'Classic' x 'Torpedo' (V6)

The experimental lots with ungrafted and grafted eggplants were set up at 27/06/2015 at greenhouse of glass and maintained according with the specific technologies (Figure 1, a and b).

The experience was made up of 96 plants/V1 or V2 (24,000 plants ungrafted/ha) and 72 plants/V3 or V4, V5 and V6 (18,000 grafted plants/ha). It was placed by the randomized block method in 4 repetitions (24 plants/ repetition V1, V2 and 18 plants/repetition V3, V4, V5 and V6).

The observations, the biometric determinations and the biochemical analyzes on eggplant fruits were performed in the laboratories of the Horting Institute on biological samples harvested to consumer maturity.

a)   After planting

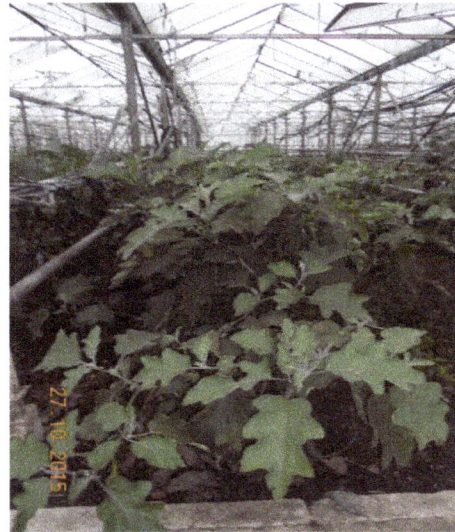

b)   During the vegetative period

Figure 1. Experimental lots with grafted eggplants cultivated into greenhouse

It were determined form index (FI), form (F), weight/fruit (W), marketable production (MP), humidity (H) – gravimetric method, soluble dry matter (SDM) by refractometry method using ABBE refractometer, total sugar (TS) by Bertrand method.

The experience was carried out under conditions of extreme temperatures, very high in the summer of 2015; these temperatures were a calamity for vegetable cultures.

RESULTS AND DISCUSSIONS

The results concerning some observations, biometric determinations and biochemical analyzes performed on eggplant fruits are shown in Table 1. The grafting did not affect the fruit form.

The grafted and ungrafted plants have had the fruits with same form, elongated eggplants (Figure 2).

Table 1. Data concerning some fruit characteristics from eggplant variants

Variants	Biometric determinations			Biochemical analyzes		
	FI / F	W (g/fruit)	MP (kg/ha)	SDM (°R)	TS (%)	H (%)
V1 (Control)	2.25 / elongate	353	50,230	4.60	1.50	92.10
V2 (Control)	2.30 / elongate	305	50,200	4.50	1.30	92.10
V3	2.25 / elongate	273	61,400	5.00	1.27	94.17
V4	2.28 / elongate	255	58,120	4.60	1.25	93.90
V5	2.27 / elongate	281	58,390	4.75	1.39	94.05
V6	2.28 / elongate	283	56,980	4.63	1.24	93.98

V1 (Control)   V2 (Control)

V3   V4

V5   V6

Figure 2. Eggplant form, elongate, V1,V2, V3, V4, V5, V6

The fruit weight was not significantly influenced by the grafting. At some grafted variants (V3 and V6), 'Emperador' and 'Torpedo' rootstocks had determined obtaining of fruits with weighing less than the ungrafted variants, control (V1 and V2) and another grafted variants (V5 and V4) had determined obtaining of fruits with weighing more bigger than the ungrafted variants, control (V1 and V2).

There is a direct linear correlation between variants and the average marketable production. The determination coefficient shows that for production (kg/ha), the correlation is very significant, $r^2 = 1$ (Figure 3).

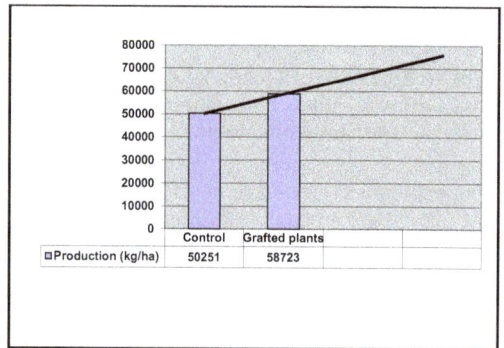

	Control	Grafted plants
□ Production (kg/ha)	50251	58723

Figure 3. Influence of grafting on eggplant production

The fruit color harvested at consumer maturity was not influenced by the grafting; such as the fruits from grafted plants had same color as the fruits from ungrafted plants, control.

The differences were insignificant about the content some biochemical components.

About SDM, the grafted variants (V2, V3, V4, V5 and V6) have had slightly higher values compared with the ungrafted variants (V1 and V2). The S content was higher in ungrafted variants (V1 and V2) compared with the grafted variants (V2, V3, V4, V5 and V6), but the differences were not significant.

Bogoescu and Doltu, 2015, show in a specialty paper that ungrafted eggplants have a higher carbohydrate content (2.54–2.97%) compared with the grafted eggplants (1.92–2.00%) cultivated under same biotope and technological conditions.

About H, the ungrafted variants (V1 and V2) has had same value and slightly lower than the grafted variants (V2, V3, V4, V5 and V6).

The results obtained by some experts (Moncada et al., 2013) about the yield and the quality of some grafted eggplants are contradictory.

CONCLUSIONS

The extreme temperatures, very high from the summer of 2015, considered a calamity, led to a strong abiotic stress on plants during the vegetable period; thus, the results about eggplant production were inconclusive.

The grafting effect on eggplants it was insignificant in the climatic conditions of 2015.

The fruits of the grafted plants have form, color and weight as the fruits than ungrafted plants.

At the grafted variants, the soluble dry matter content was higher, the sugar content was lower and the humidity was higher than in the ungrafted variants, but the differences were not significant.

The rootstock and scion plants are compatible regarding the fruit quality.

The results obtained by the research in 2015 concerning the grafting effect on the eggplant production are not relevant; thus it requires further researches on these issues.

The conclusions of the researchers from the Horting Institute are in keeping with some conclusions of the foreign researchers, but are required more researches in this domain for to highlight the grafting effect on some aspects concerning production of grafted vegetables.

ACKNOWLEDGEMENTS

This research was conducted within the internal thematic plan of the Horting Institute Bucharest from 2015, project supported from the funds of the institute.

REFERENCES

Bogoescu M., Doltu M., Sora D., Iordache B., 2008. Results on establishing the technology for obtaining the tomatoes grafted seedlings designed for greenhouses. Bulletin UASVM Horticulture Cluj-Napoca, vol. 65(1), 147-152.

Bogoescu M., Doltu Mădălina, 2015. Effect of Grafting Eggplant (*Solanum melongena* L.) on its Selected Useful Characters. Bulletin UASVM Horticulture Cluj-Napoca, vol. 72(2), 313-317.

Çürük S., Dasgan H. S., Mansuroglu S., Kurt S., Mazmanoglu M., Antaklı O., Tarla G., 2009. Grafted eggplant yield, quality and growth in infested soil with *Verticillium dahliae* and *Meloidogyne incognita*. Pesq. agropec. bras., Brasília, vol. 44(12), 1673-1681.

Davis A.R., Perkins-Veazie P., Hassell R., King S.R., Zhang X., 2008. Grafting effects on vegetable quality. HortScience vol. 43, 1670-1672.

Doltu M., 2007. Stabilirea tehnologiei pentru producerea de rasadurilor altoite de tomate. Editura Agris Bucuresti, Revista Horticultura nr. 9, 4.

Echevarria P.H., Rodriguez A., Vivaracho S.M., Vallejo A.D., 2004. Influence of rootstocks and soil treatment on the yield and qulity of greenhouse-grown cucumbers in Spain – Acta Horticulture vol. 663, 403-408.

King S.R., Davis A.R., Zhang X., Crosby K., 2010. Genetics, breeding and selection of rootstocks for Solanaceae and Cucurbitaceae. Sci. Hort. vol.127, 106-111.

Lee J.M., 1994. Cultivation of grafted vegetables I: current status, grafting methods and benefits. Hortscience vol. 29, 235-239.

Moncada A., Miceli A., Vetrano F., Mineo V., Planeta D., D'Anna F., 2013. Effect of grafting on yield and quality of eggplant (*Solanum melongena* L.). Scientia Horticulturae, Advances in Vegetable Grafting Research vol. 149, 108-114.

20

GOOD AGRICULTURAL PRACTICES (GAP) FOR GREENHOUSE SOILLESS TOMATO GROWING

Nilda ERSOY[1], Osman TEKİNARSLAN[2], Ulaş GÖKTAŞ[3], Elif AKÇAY ÖZGÜR[2], Ayşe METİN[4], Ramazan GÖKTÜRK[5], Gürkan KAYA[6]

[1]Akdeniz University, Vocational School of Technical Sciences, Program of Organic Agriculture, 07058, Antalya, Turkey

[2]ECAS Inspection and Certification Organization, 07075, Antalya, Turkey

[3]PROANALİZ Laboratories Group, Antalya, Turkey

[4]ESAY Agriculture Consulting, 07310, Antalya, Turkey

[5]Akdeniz University, Faculty of Science, Department of Biology, 07058, Antalya, Turkey

[6]Mehm et Akif Ersoy University, Faculty of Engineering and Architecture, 15030, Burdur, Turkey

Corresponding author email: nildaersoy@akdeniz.edu.tr

Abstract

Tomatoes have become one of the most popular and widely grown vegetables around the world. Nowadays, tomato cultivation with soilless agriculture has gained popularity. Sustainability of good farming practices seems to be quite high compared to other agricultural production systems. This research was carried out in the Serik District of Antalya to demonstrate the effectiveness of good agriculture practices in the greenhouse cultivation of soilless tomatoes. The varieties taken in the experiment were Gülpembe grafted on Beaufort rootstock. In the study, pesticide residue levels were determined in fruit extracts as well as analyses for water. Extraction steps and all analyses were carried out at the Proanaliz Food Control Laboratory. High-precision analytical instruments such as LC-MS / MS and GC-MS were used for pesticides residues. A total of 1025 pesticide active substances were analysed in LC-MS / MS and 117 pesticide active substances in GC-MS in fruit extracts. In this research carried out in 2015 and 2016, samples of both years were not found to be detectable to the tolerance values of Turkish Food Codex (TFC).

Key words: tomatoes, GAP, pesticides, residue, Serik-Antalya.

INTRODUCTION

Tomato is one of the most important vegetables produced in the world. Its homeland is the South America Countries where Peru and Ecuador are located. It was first cultivated by Mexicans and after the discovery of New World, it spread from America to Europe and the other continents. In Turkey, it was started to be grown in the early 1900s in Adana.

On the species basis, the tomato that takes the first place in vegetable cultivation is cultivated both open field and in greenhouses.

In Turkey 12,615,000 tons of tomatoes are produced, of which 8,170,000 tons are table tomatoes, 4,445,000 tons are for sauce production (TÜİK, 2016).

Greenhouse production has an important place in our agricultural sector due to its high efficiency and income from the unit side and ,at the same time, providing a regular use of

labor force throughout the year by spreading vegetable production to all seasons of the year. In greenhouse production, tomato takes the first place.

Approximately 27% (3,399,100 tons) of our total tomato production is obtained from the greenhouse and contribution of this specie in greenhouse production is 53.5%. 77.6% of greenhouse tomato production is originating from the greenhouse systems in the Mediterranean Region.

Antalya provides 62.5% of this production. Greenhouse tomato production is practiced in 259,709 da areas. 5.6% of production area consists of low plastic tunnel (14,644 da), 6.1% high plastic tunnel (15,765 da), 68.5% plastic greenhouse (177,937 da) and 19.8% glass greenhouse areas (51,363 da). Most of the production is obtained from plastic greenhouses (71%) and glasshouses (21%).

The amount of greenhouse tomato production in Turkey has increased prominently in recent years due to increase of production area, usage of quality seed and modern agricultural techniques. (TÜİK, 2016).

Soilless production in Turkey besides having a history of about 20 years, is rapidly becoming widespread due to providing earliness, productivity and quality increase.

Soilless agriculture is generally widespread in the Mediterranean and Aegean Regions and 74% of total production is made in Antalya and İzmir provinces.

The most important specie in soilless agriculture is tomato (Toprak and Gül, 2013; Kandemir et al., 2015).

Soilless agriculture is based on the principle that water and nutrients required for plant growth are supplied to the root in the required amount and it is split up into environmental and aquatic culture.

In environmental culture, plants are grown in organic or inorganic environments with irrigation nutrition solution. In aquatic culture, plants are grown in water containing nutrients. Food safety and quality are among the most important issues in recent years.

The residues of the pesticides have become the fear of those who wants healthy foods. The consumption of pesticides in the Mediterranean and Aegean regions, where greenhouses are common, is close to two-thirds of the country's total.

On the other hand, when we think features of pesticides we consume, the vast majority have significant risks for human and environmental health.

Residue analyzes which are made more frequently than the past show that pesticide contamination in our products is reduced, but even in our elite products exported to EU countries, we encounter parcels which are not suitable for pesticide residue limits.

Besides all these problems, a number of new legal regulations to solve problems caused by pesticides, the prohibition of some risky pesticides, the introduction of a prescription system, and some other planned regulatory developments are seen as promising developments (Anonymous, 2017a).

Another of these promising developments, which has a great importance, is undoubtedly good agricultural practices based on certificated production. Antalya has an important position in vegetable, especially tomato growing. In such places where the potential is high, it is necessary to speed up the works for good agricultural practices.

In this research, it is aimed to determine pesticide residues obtained from a greenhouse, in which are grown tomatoes with good agricultural practices, in hydroponic system in Antalya.

MATERIALS AND METHODS

Materials

Materials in this research are varieties of truss tomato called 'Gülpembe F1' grafted onto rootstock. This kind of plant is in the category of pitted tomato, the shortness of internode is short, early and highly productive. There are 4-5 fruits in the bunch, they are small sliced, their taste and aroma are so good, colour is bright pink, weight 250 g, flesh is hard and its shelf life is ideal.

Pesticides given in Table 1 and Table 2 are searched in the examples of tomatoes, which are the materials. All extraction studies and residue analysis of the examples made in Proanaliz Food Control Laboratory.

Table 1.Active substances examined in tomatoes examples on LC-MS/MS device

2,4,5-T; 2,4-D; 3,4,5 trimethacarb; Abamectin; Acephate; Acequinocyl; Acetamiprid; Acetochlor; Acibenzolar-S-Methyl; Aclonifen; Acrinathrin; Alachlor; Aldicarb; Aldicarb-Sulfone; Aldicarb-Sulfoxide; Allethrin; Ametoctradin; Ametryn; Amidosulfuron; Amisulbrom; Amitraz; Amitrole; Anilazine; Anilofos; Aramite; Asulam; Atrazine; Azamethiphos; Azimsulfuron; Azinphos-Ethyl; Azinphos-Methyl; Aziprotryne; Azoconazole; Azocyclotin; Azoxystrobin; Barban; Beflubutamid; Benalaxyl; Bendiocarb; Benfuracarb; Benomyl; Bensulfuron methyl; Bentazone; Benthiovalicarb Isopropyl; Benzoximate; Bifenox; Bifentrin; Binapacryl; Bioresmethrin; Bispyribac; Bitertanol; Boscalid; Bromacil; Bromophos methyl; Bromophos-Ethyl; Bromoxynl; Bromuconazole; Bupirimate; Buprofezine; Butafenacil; Butocarboxim; Butocarboxim-sulfone; Butocarboxim-sulfoxide; Butoxycarboxim; Butralin; Buturon; Butylate; Cadusafos; Campheclor; Campheclor-methyl; Campheclor-oxon; Campheclor-oxon-sulfone; Campheclor-oxon-sulfoxide; Campheclor-sulfone; Campheclor-sulfoxide; Carbaryl; Carbendazim; Carbofuran; Carbosulfan; Carboxin; Carfentrazone-Ethyl; Chlorantraniliprole; Chlorbromuron; Chlorbufam; Chlorfenvinphos; Chlorfluazuron; Chloridazon; Chlormequat chloride; Chlorotoluron; Chloroxuron; Chlorpropham; Chlorpyrifos; Chlorpyrifos-Methyl;

Chlorsulfuron; Chlortal-dimethyl; Chlorthiamid; Chromafenozide, Cinidon-ethyl; Clethodim; Clethodim Iminsulfone; Clethodim Iminsulfoxide; Clethodim Sulfoxide; Climbazole; Clodinafop-proparpyl ester; Clofentezine; Clomazone; Cloquintocet-methylhexyl ester; Clothianidin; Coumaphos; Crimidine; Cyanazine; Cyanofenphos; Cyazofamid; Cyclanilide; Cycloate; Cycloloxydim; Cyflufenamid; Cyhalofop; Cyhalofop butyl; Cyhalofop diacid; Cyhexatin; Cymoxanil; Cyproconazole; Cyprodinil; Cyromazine; Daminozide; Dazomet; Demeton(O+S); Demeton-S-Methyl; Demeton-S-Methyl-Sulfone; Demeton-S-Methyl-Sulfoxide; Desmedipham; Desmetryn; Diafenthiuron; Dialifos; Di-Allate; Diazinon; Dichlofenthion; Dichlofluanid; Dichlorprop; Dichlorvos (DDVP); Diclobutrazol; Diclofop-Methyl; Dicloran; Dicrotophos; Diethofencarb; Difenoconazole; Diflubenzuron; Diflufenican; Dimefox; Dimethachlor; Dimethenamid; Dimethoate; Dimethomorph; Dimetilan; Dimoxystrobin; Diniconazole; Dinitramine; Dinocap; Dinoseb; Dinoterb; Dioxacarb; Diphenamid; Dipropetryn; Disulfoton; Disulfoton Sulfone; Disulfoton Sulfoxide; Ditalimfos; Dithianon; Diuron; DNOC; Dodine; E-Fenpyroxymate; Emamectin; Benzoate; Endosulfan Sulfate; Epichlorohydrin; EPN; Epoxiconazole; EPTC; Esfenvalerate; Etaconazole; Ethalfluralin; Ethametsulfuron Methyl; Ethiofencarb; Ethiofencarb-sulfone; Ethiofencarb-sulfoxide; Ethion; Ethiprole; Ethirimol; Ethofenprox; Ethofumesate; Ethoprophos; Ethoxyquin; Ethoxysulfuron; Ethylene thiourea; Etoxazole, Etridiazole; Etrimfos; Famoxadone; Famphur, Fenamidone; Fenamiphos; Fenarimol; Fenazaquin; Fenbuconazole; Fenbutatin oxide; Fenhexamid; Fenitrothion; Fenobucarb; Fenoxyaprop-P-ethyl; Fenoxycarb; Fenpiclonil; Fenpropathrin; Fenpropidin; Fenpropimorph; Fensulfothion; Fenthion; Fenthion Oxon; Fenthion Oxon Sulfone; Fenthion Oxon Sulfoxide; Fenthion-Sulfone; Fenthion-Sulfoxide; Fentin acetate; Fentin Hydroxide; Fenvalerate; Fipronil; Flamprop-M-Isopropyl; Flazasulfuron; Flonicamid; Florasulam; Fluazifop-p-butyl; Fluazinam; Flubendiamide; Flubenzimine; Flucycloxuron; Flucythrinate; Fludioxonil; Flufenacet; Flufenoxuron; Flumioxazine; Fluometuron; Fluopicolide; Fluopyram; Fluorochloridone; Fluoroglycofen Ethyl; Fluoxastrobin; Flupyrsulfuron Methyl; Fluquinconazole; Fluroxypyr; Flurtamone; Flusilazole; Flutolanil; Fluxapyroxad; Fomesafen; Fonofos; Foramsulfuron; Forchlorfenuron; Formetanate; Formetanate hydrochloride; Fosthiazate; Fuberidazole; Furalaxyl; Furathiocarb; Halfenprox; Halosulfuron Methyl; Haloxyfop-2-Ethoxy-Ethyl; Heptanafos; Hexaconazole; Hexaflumuron; Hexazinone; Hexythiazox; Imazalil; Imazamox; Imazapic; Imazapyr; Imazaquin; Imazethapyr; Imazosulfuron, Imibenconazole; Imidachloprid; Indoxacarb; Iodosulfuron methyl sodium; Ioxynil; Ipconazole; Iprobenfos; Iprodione; Iprovalicarb; Isazofos; Isocarbofos; Isoprocarb; Isoproturon; Isoxaben; Isoxadifen Ethyl; Isoxaflutole; Isoxathion; Kinetin; Kresoxim-methyl; Lenacil; Linuron; Lufenuron; Malaoxon; Malathion; Maleic Hydrazide; Mandipropamide; MCPA; Mecarbam; Mecoprop (MCPP); Mecoprop-P (MCPP-P); Mepanipyrim; Mephosfolan; Mepronil; Meptyldinocap; Mesosulfuron Methyl; Mesotrione; Metaflumizone; Metalaxyl; Metalaxyl M; Metamitron; Metazachlor; Metconazole; Methabenzthiazuron; Methacrifos; Methamidophos; Methidathion; Methiocarb; Methiocarb sulfone; Methiocarb sulfoxide; Methomyl; Methomyl oxime; Methomyl Sulfone; Methoxyfenozide; Metobromuron; Metolachlor; Metolcarb; Metosulam; Metoxuron; Metribuzin; Metrofenone; Metsulfuron-Methyl; Mevinphos; Milbemectin A3; Milbemectin A4; Molinate; Monocrotophos; Monolinuron; Monuron; Myclobutanil; Naled (Dibrom); Naphthalene Acetamide (NAD); Napropamide; Napthol-1; Neburon; Nicosulfuron; Nitenpyram; Norfluazuron; Novaluron; Nuarimol; Ofurace; Omethoate; Orthosulfamuron; Oxadiazon; Oxadixyl; Oxamyl; Oxasulfuron; Oxycarboxin; Oxyfluorfen; Paclobutrazol; Paraoxon Ethyl; Paraoxon Methyl; Parathion-Ethyl; Parathion-Methyl; Pebulate; Penconazole; Pencycuron; Pendimethalin; Penoxsulam; Permethrin; Pethoxamid; Phenmedipham; Phenothrin; Phentoate; Phorate; Phorate Sulfone; Phosalone; Phosmet; Phosmet oxon; Phosphamidon; Phoxim; Picolinafen; Picoxystrobin; Pinoxaden; Pirimicarb; Pirimicarb Desmethyl; Pirimicarb Desmethyl Formamido; Pirimiphos-Ethyl; Pirimiphos-Methyl; Prochloraz; Profenofos; Profoxydim; Profoxydim lithium; Prohexadione calcium; Promecarb; Promethryn; Propachlor; Propanil; Propaquizafop; Propargite; Propazine; Propetamphos; Propham; Propiconazole; Propisochlor; Propoxur; Propoxycarbazone sodium; Propyzamide; Proquinazid; Prosulfocarb; Prosulfuron; Prothioconazole; Prothiophos; Pymetrozine; Pyraclostrobin; Pyraflufen; Pyraflufen ethyl; Pyrasulfotole; Pyrazophos; Pyrethrins; Pyridaben; Pyridaly; Pyridaphenthion; Pyridate; Pyrifenox; Pyrimethanil; Pyriproxyfen; Quaizalofop_P_Ethyl; Quinalphos; Quinclorac; Quinmerac; Quinoxyfen; Resmethrin; Rimsulfuron; Rotenone; Sethoxydim; Silthiofam; Simazine; Spinetoram; Spinosad; Spirodiclofen; Spiromesifen; Spirotetramat; Spirotetramat-Enol; Spirotetramat-Enol-Glucoside; Spirotetramat-Ketohydroxy; Spirotetramat-Monohydroxy; Spiroxamine; Sulcotrione; Sulfosulfuron, Sulfotep; Sulprofos; Tebuconazole; Tebufenozide; Tebufenpyrad; Tebupirimfos; Teflubenzuron; Tembotrione; Temephos; TEPP(O.O-TEPP); Tepraloxydim; Terbufos; Terbumeton; Terbuthylazine; Terbutryn; Tetraconazole; Tetramethrin; Thiabendazole; Thiacloprid; Thiamethoxam; Thidiazuron; Thifensulfuron-methyl; Thiobencarb; Thiodicarb; Thiofanox; Thiofanox Sulfone; Thiofanox Sulfoxide; Thiophanate-methyl; Tolclofos-Methyl; Tolfenpyrad; Topramezone; Tralkoxydim; Triadimefon; Tri-allate; Triasulfuron; Triazophos; Tribenuron-Methyl; Trichlorfon; Trichloronat; Triclopyr; Tricyclazole; Tridemorph; Triethyl Phosphate; Trifloxystrobin; Triflumizole; Triflumuron; Triflusulfuron Methyl; Triforine; Trinexapac Ethyl; Triticonazole; Tritosulfuron; Uniconazole; Vamidothion; Zoxamide

Table 2.Active substances examined in tomatoes examples on GC-MSD device

2,4-5T; 4,4 Dichlorobenzophenone; Aldrin (HHDN); Alpha-Endosulfan; Alphamethrin (cypermethrin); Beta-Endosulfan; Bromopropylate; Captafol; Captan; Chlorfenson; Chlorobenzilate; Chlorothalonil; Cycloate; Cyfluthrin; Cypermethrin; Dazomet; DDD-2.4'-; DDD-4.4'-; DDE-2.4'-; DDE-4.4'-; DDT-2.4'-; DDT-4.4'-; Deltamethrin; Dicamba; Dicofol; Dinobuton; Diphenylamine; Endosulfan-sulfate; Endrin; Esfenvalerate; Ethalfluralin; Fenvelarate; Folpet; Formothion; Gamma-HCH (Lindane); Hexachlorobenzene; Iprodione; IS-TPP; Lambda-Cyhalothrin;

Methoxychlor; Nitrofen; Nitrothal-isopropyl; Permethrin; Phenmedipham; Procymidone; Quintozene (penta chloro nitro benzene); Qunomethionate; Tau-fluvalinate; Tecnazene; Tetradiphon; Tetrasul; Thiometon; Tolyfluanid; Trifluralin; Vinclozolin; α (alpha)- HCH; β(beta) – HCH; 2-chloraniline; 3-chloraniline; 4-chloraniline; Aminocarb; Diclobenil; Biphenyl; Propamocarb; Carbofuran -3-hydroxy; Nitrapyrin; Chloroneb; 2-phenyl phenol; Benfluralin; BHC; Dioxathion; Profluralin; Fluchloralin; Terbacil; HCH(delta); Tefluthrin; Bromociclen; Pentacholoaniline; Flurprimidol; Chlorzolinate; Transfluthrın; Fenchlorphos; Diphenylmercury; Dinoseb acetate; Heptachlor endo-epoxide(trans isomer); Heptachlor endo-epoxide(cis isomer); S-Metolachlor; Fenson; Isodrin; Iodofenphos; Isofenphos; Chlorbenside; Haloxyfop-R-methyl; Tetrachlorvinphos; Chlordane-cis-alpha; Flutriafol; Diethatyl-ethyl; Dieldrin; Perthane; Chlorfenapyr; Carbophenothion, Tributyl Phosphote; Chlordane-trans-gamma; IS-TPP; Bifenazate; Mefenpyr-diethyl; Leptophos; Heptachlor; Mirex; Dimethipin; Cyanaphos; Chlorthion; Methoprene; Chlordecone; Oxadiargly; Fluotrimazole; Lactofen

Methods

All the solvents and chemicals (water, acetonitrile, methanol, formic acid, acetic acid and ammonium formate) used as mobile phases in example extractions are chosen in accordance to a profound quality. Pesticide standards are prepared at least a 90% rate of purity. Extractions and clearance of the examples are generalized in accordance with AOAC (International Official Methods of Analysis) methods (Lehotay, 2007). Some chromatographic conditions for LC-MS /MS and GC-MS devices are given in Table 3 and Table 4.

Table 3. Chromatographic Working Conditions of LC-MS/MS

LC-MS/MS	Agilent 6420			
Mobile Phase A	5 mM Amonium Formate&Water + Acetonitrile			
Mobile Phase A	Pure methanol			
Column	Poroshell 120 SB-C18 (3.0 x 100 mm 2.7 Micron)			
Injection Volume	10 µl			
Flow Rate	0.6 ml/min			
MS Gas Temperature	300°C			
Sheat Gas Temperature	350°C			
The Column Oven	35°C			
Pump Gradient Program				
	Time	Mobile phase A %	Mobile phase B %	Flow rate ml/min
	0:00	80	20	0.6
	0:00	80	20	0.6
	0:20	80	20	0.6
	1:50	30	70	0.6
	6:00	5	95	0.6
	7:50	5	95	0.6
	7:60	80	20	0.6
	10:00	80	20	0.6

Table 4. Chromatographic Working Conditions of GC/MS

GC-MS	Agilent 5975		
Carrier gases	Helium		
Column	HP-5MS 30 m × 250 µm × 250 µm × 0.25 µm		
Injection Volume	5 µl		
Flow Rate	2.4 ml/min		
Duration of Injection	18.5 min		
MS Gas Temperature	300°C		
Sheat Gas Temperature	350°C		
The Column Oven	35°C		
Inlet temperature program			
Start	Rate of increase (°C/min)	Temperature (°C)	Retention Time (RT) (min)
1	0	55	0.21
2	600	325	18.5

The Column Oven temperature program			
Start	Rate of increase (°C/min)	Temperature (°C)	Retention Time (RT) (min)
1	0	50	0
2	50	150	0
3	20	230	1
4	8	290	3
5	0	290	18.5

RESULTS AND DISCUSSIONS

Residue quantities obtained from the research were evaluated as average of 3 repetitions in each sample according to Turkish Food Codex (TFC) Regulation on Maximum Residue Limits of Pesticides (Turkish Official Gazette No 21.01.2011-27822; Notification No: 2011/2). The TFC residue limits of each pesticide sample are indicated separately in the tables presented. In residue limits determined by using high-precision analytical instruments such as GC-MS and LC-MS/MS, in tomatoes samples analyses of total 1025 pesticide active ingredients were made in LC-MS/MS instrument and 117 pesticide active ingredients in GC-MS instrument. In this research carried out between 2015 and 2016, detectable levels of the residues were not found in the samples of these two years.

The use of conscious and controlled pesticides in our agricultural production, especially in exported products, is very important to avoid the residual problem. The use of pesticides should be very conscious and controlled to ensure the safety of our country's people and to protect our environment and our foreign trade. In the EU and the US, priority should be given to low-risk or environmentally friendly pesticides that have the potential to affect the environment and health as little as possible. In Turkey, environmentally friendly pesticides are not given priority to licensing and supporting of their consumption (Tiryaki, 2016).

Not having pesticide residues of agricultural products is very important in domestic consumption and foreign trade. Because the communication is very fast. In Rapid Alert System for Food and Feed (RASFF) of the EU is notified of which are not suitable due to pesticide residues. The EU has published products and origins on the internet which are not suitable for residues in the EU products through the Rapid Alarm System (Anonymous, 2017b). The eligibility status of the products exported to EU countries is shown in Table 5.

As seen in Table 5, the number of lots that are not suitable for the standards of foods and feedstuffs sent to the EU from Turkey and it rose even further in 2015.

Turkey is in the second place among 146 countries from the point of the number of non-eligible parties. Also, in a research we carried out, 203 pesticide residues in tomato, pepper and aubergine vegetables samples collected from local markets and markets in Konya were analyzed.

Extraction of vegetable samples was carried out in the laboratories of Selçuk University Departments of Agronomy and Horticulture, residue analyzes were made in the laboratories of Ministry of Food, Agriculture and Livestock, İzmir Provincial Control Laboratory Directorate's Organic Farming Products and Residue Analysis Laboratory with LC-MS/MS and GC-MS devices. Findings obtained from the study show that oxamyl (Tolerance value of Turkish Food Codex (TFC); 10 μg/kg), which is totally forbidden to use in a tomato sample, has a value of about 7 times, two different pesticides (112 μg/kg Ethion and 75 μg/kg Triazophos) were found in a pepper sample , in another pepper case, it was determined that 120 μg/kg Benomyl was above the tolerance value of 100 μg/kg of TFC.

In the 10 aubergine samples taken for the experiment, it was determined that the level of oxamyl which is prohibited to use was about 11 times, which means, 107 μg/kg. Besides, Imidiacloprid (TFC tolerance value; 20 μg/kg) was found respectively at 49, 190 and 64 μg/kg in 3 different aubergines.

Table 5. 2013-2015 notifications by country of origin

Country	Year 2013	Year 2014	Year 2015
China	436	417	388
Turkey	226	200	282
India	257	199	276
Spain	185	169	159
France	120	104	120
Poland	164	131	118
Germany	95	135	117
Italy	105	89	117
Netherlands	103	114	94
Brazil	187	109	91
United States	102	164	87
Vietnam	76	124	85
Egypt	49	55	78
Thailand	88	90	71
Iran	21	54	61
Belgium	60	75	58
United Kingdom	55	50	56
Denmark	19	28	27
Sweden	45	7	25
Argentina	76	40	22

In the same way, Osman et al. (2010) also investigated 23 different pesticide residues from GC-MS in 160 local vegetables collected from 4 large supermarkets in Al-Qassim district of Saudi Arabia (2010). According to the results, in 89 of 160 samples were found pesticide residues, in 53 it was determined that obtained values was above the Maximum Residue Levels (MRL). In this research, pesticide residues was found in 17 of 30 vegetable samples.

The most common pesticides in vegetables were respectively Carbaryl, Biphenyl and Carbofuran.

Zengin and Karaca (2017), investigated on 249 different pesticides residue levels in tomatoes samples which taken from greenhouse in Uşak province in 2015-2016 growing seasons. According to the results, 63% of taken 60 tomato samples had no detectable residues. In 37% of tomato samples detected pesticide residue, none of this pesticides exceeded the maximum residue limits given in Turkish Food Codex. Imidacloprid was the most common pesticide among detected pesticides.

Pesticide residues obtained from 11 of 16 pumpkin samples, 7 of 12 carrot and green pepper samples, 6 of 11 cucumber samples, 5 of 12 eggplant samples, 7 of 11 spinach samples, 6 of 11 lettuce samples, 4 of 11

tomato samples were reported to be above MRL.

The highest pesticide residues were found respectively in lettuce (Ethiofencarb, 7.648 mg/kg), tomato (Tolclofos-methyl, 7.312 mg/kg), pumpkin (Chlropyrifos, 6.207 mg/kg), carrot (Heptanophos, 3.267 mg/kg), green pepper (Carbaryl, 2.228 mg/kg) and aubergine (Carbaryl, 1.917 mg/kg). These findings point that it's necesssary pesticide residues in vegetables grown in greenhouses to be examined for the protection of public health.

Duru et al. (2013), examined pesticide residues in 33 of 145 fruit and vegetable samples which are sampled in İzmir in 2007-2008. In 74 of 145 analyzed samples was not found residue. In 30 samples were found residues below tolerance, in only 7 samples were found residues over-tolerance. The highest residues were found in the fresh grape, raisins and tomato samples.

Tatlı (2006) investigated the residue levels of pesticides commonly used in fresh fruit and vegetable samples (strawberry, tomatoes, artichoke, table figs, cherry, potatoes, peach, table grapes, olive) collected in the Aegean region and from human consumption areas and in the cultivation of dried food samples. In these samples, 50 pesticides with organic chlorine, organophosphorus insecticides and synthetic pyrethroids, strobulin and benzimidazole group were selected from fungicides. In tomato, artichoke, table figs, dried figs and potato samples were not found any detectable pesticide residues. And in the samples of other products at least one pesticide residue was found at detectable levels.

Residue quantities in the samples with residuals were evaluated according to Turkish Food Codex and EU MRLs and pesticide residues were found over 2.34% tolerance in agricultural products.

CONCLUSIONS

The usage of pesticides and the fate of this usage are always on the agenda in today's world, and it is also likely to remain like that. Because, as intensive pesticide use in traditional agriculture, there are controlled pesticide use in "Good Agricultural Practices" and natural pesticide use in organic agriculture. In many studies, toxicological risks and environmental risks of pesticide residues to human health have been researched as a result of excessive and unconscious pesticide use (Tiryaki, 2016). In this context, Good Agricultural Practices involves agricultural techniques which are environmentally sensitive, do not harm human and animal health, aim to protect natural resources, and ensure traceability and food safety.

With such production techniques, socially viable, economically profitable and sustainable agricultural production is targeted. Therefore, Turkey has the support of the Ministry of Food, Agriculture and Livestock to encourage the transition to good agricultural practices. When compared to conventional agriculture, these can be promising techniques for achieving healthy and reliable food.

REFERENCES

Anonymous 2017a.
    http://www.zmo.org.tr/resimler/ekler/52cf38361a20
    908_ek.pdf (accessed, March 25 2017).
Anonymous 2017b.
    https://ec.europa.eu/food/sites/food/files/safety/docs
    /rasff_annual_report_2015.pdf (accessed, March 25
    2017).
Duru A.U., Örnek H., Burçak A.A., 2013. İzmir İli'nde
    satışa sunulan bazı meyve ve sebzelerde pestisit
    kalıntılarının araştırılması. I. Bitki Koruma Ürünleri
    ve Makinaları Kongresi, Bitki Koruma Ürünleri
    (Cilt I, Pestisitler), Bildiriler Kitabı: 59-69, 2-5.
Kandemir D., Kurtar E.S., Demirsoy M., 2015. Türkiye
    Sebze Fidesi Üretimindeki Son Gelişmeler,
    TÜRKTOB Türkiye Tohumcular Birliği Dergisi,
    No:4, 4.
Lehotay S.J., 2007. Determination of pesticide residues
    in foods by acetonitrile extraction and partitioning
    with magnesium sulfate: collaborative study,
    Journal of AOAC International, 2007 Mar-Apr,
    90(2):485-520.
Tiryaki O., 2016. Türkiye'de yapılan pestisit kalıntı
    analiz ve çalışmaları, Erciyes Üniversitesi Fen
    Bilimleri Enstitüsü Dergisi, 32(1):72-82.
Toprak E., Gül A., 2013. Topraksız tarımda kullanılan
    ortam domates verimi ve kalitesini etkiliyor mu?,
    Tarım Bilimleri Araştırma Dergisi, 6 (2): 41-47.
TÜİK, 2016. Türkiye İstatistik Kurumu,
    www.tuik.gov.tr/ (accessed, March 25 2017).
Zengin E. and Karaca İ. 2017. Uşak İlinde Örtü Altı
    Üretimi Yapılan Domateslerdeki Pestisit
    Kalıntılarının Belirlenmesi, Süleyman Demirel
    University Journal of Natural and Applied
    Sciences, Published Online: 03.02.2017.

21

GREEN ROOF VEGETATION POSSIBILITIES

Zuzana POÓROVÁ, Zuzana VRANAYOVÁ

Technical University of Košice, Civil Engineering Faculty, Vysokoškolská 4 Košice 042 00, Slovakia, Em ail: zuzana.poorova@tuke.sk, zuzana.vranayova@tuke.sk
Corresponding author email: zuzana.poorova@tuke.sk

Abstract

The aim of this article is showing few ways how to make green roofs. The difference between intensive and extensive green roof, plant establishment and choice of plants is described. Aim of this article is showing the best combination of designed roof construction, designed thickness of soil and designed vegetation. Types of soil, depending on intensive or extensive green roof are described. Types of vegetation, depending on intensive or extensive green roof are described.

Key words: establishment, green roof, medium, plant.

INTRODUCTION

In just few years, green roofs have gone from a historical curiosity to a booming growth industry – primarily because the environmental benefits of extensively planted roofs are now beyond dispute, whether for industrial or governmental complexes or private homes in urban or suburban settings. This paper deals with extensive green roofs. Nature helping the body, views offered for people living near green roofs, relaxant attributes of wildness in built environment. In bigger detail it focuses on its plants and vegetation, its construction, attributes and best use and choice of plants.

Medium depth and its greater depth means more diversity of used plants because of more options for growing roots of used plants. Composition of the underlying medium influences load of soil. This also means influencing plant specification in terms of weight, water absorption capacity, drainage rates etc. The ideal medium is lightweight, retaining water well, also porous and freely draining. The more water the medium retains, the more weight is being added to the roof. The medium supplies and absorbs nutrients, anchors the plants, provides enough weight to avoid floating when wet and avoids being flown off during establishment (Dunett, 2004).

Extensive green roofs are lightweight veneer systems of thin soil or substrate layers of drought tolerant self-seeding vegetated roof covers. Extensive green roofs require special types of plants. Also, they can be constructed on existing structures with little, or no additional structural support.

Generally, extensive green roof medium is a blend of sandy or granular materials that balances water absorption with adequate porous surface. A variety of natural and unnatural materials can be used to achieve balance. Lelite, pumice, diatomaceous earth, sand, expanded and active clays, expanded shale, gravel, bricks and tiles. And vermiculite or perlit can be used in conjunction with other materials (Snodgrass, 2006). But we need to face the fact that using these kinds of materials the green roof is going to be less environmental and more expensive than purely natural medium.

Intensive green roofs are designed to look like gardens, landscapes. They need similar management as ground gardens. Contemporary technological conditions allow many things. Waterproof membranes help to capture water for irrigation, drainage support growing medium and resist invasion of roots of plants. During the day, temperature of asphalt roof is unbelievably high. On green roof, soil mixture and vegetation act like an insulation. Reducing heating; cooling the building. When it is raining, water floods down to city´s artificial canyons. A living roof absorbs water, filters it and slows it down.

More organic medium, more planting options are available. Predominantly organic medium is not recommended for extensive green roofs.

Because of decreasing of pore space, higher water retention and increasing nutrient loading, reducing medium depth over time may be caused. Changing of medium depth may cause change of the designed roof, adding the substrate and changing environment of planted vegetation. Depth of medium should be constant over a long period of time and highly organic medium makes it impossible.

MATERIALS AND METHODS

Plant establishment is the key to green roof´s longevity. If the establishment in the beginning is unsuccessful, time of the return of investments is going to be lengthened. It is very important and also much cheaper to ensure the plant establishment in the beginning or even before the realization of the roof. First weeks after installation are crucial. It is prudent to plant the plants early enough to allow plants to root in before the first frost. Trials performed at Penn State University on plant establishment showed that well-established plants were much more likely to survive winter and drought than plants that were poorly established (Thuring, 2014).

Proper care during establishment will provide achieving coverage in earlier date. Planting occurs regular irrigation. If planting occurs in areas with natural rainfall that is regular, irrigation may not be needed. On many installations on US East Coast, plants require no supplemental irrigation at all, not even upon planting. On the other hand, parts in North require care and every day irrigation. Irrigation can be achieved through several methods: built in irrigation systems, lawns sprinklers, garden hoses. Irrigation need should be ascertained and used for the specific plants, location and time of year when the roof is being installed (Snodgrass, 2006).

Seeds are first way of installing green roof. No wholly seeded green roof installations exist in North America, but it seems likely that they will eventually appear. Market pressures to decrease installation costs by direct sowing on green roofs that could become more viable and the least expensive method (Snodgrass, 2006).

Seeded green roof takes the most time to mature, generally two to three years for coverage. Limited numbers of species can reliably germinate on a roof. All require some supplementary irrigation during germination and establishment phase. Seeds are best sown in spring or fall, depending on climate. To achieve full coverage of a roof in a short time period, quicker maturing annuals could be mixed with perennial seeds.

Cuttings are the most used plants installed on green roofs. They are viable and increasingly popular method for establishing. They are quicker than seeds and depending on climate, place and time, they may not need any supplementary irrigation to help them to establish. Cuttings are more expansive than seeds, but they achieve coverage much earlier. They can cover the roof within a year after planting (Nakano et al. 2014).

Plugs are cuttings with established root system. They offer a compromise between cost and flexibility. They offer greater diversity, because fully rooted plugs store sufficient energy to allow for easy establishment. They are easily packed in boxes (Nakano et al. 2014).

Nursery containers are occasionally specified for extensive green roofs when more established plants are needed from the beginning. Where the medium is deep enough to accommodate the root system, vegetation will spread more quickly than of plugs. If the depth of the root ball is bigger than depth of soil, root ball needs to be broken, roots need to be shortened to fit into soil (Nakano et al. 2014).

Vegetative mats are long rolls of pregrown plants set in a thin layer of mesh and medium. They are fully mature upon installation. They are installed in strips on top of a base substrate, which provides eventual root support. Mats are heavy and bulky to transport, must be grown at least one year before installation (Snodgrass, 2006).

Modules are discrete vegetative systems of black plastic squares or rectangles. They are the most expensive green roof planting option. They share all advantages and disadvantages of mats, but they include more medium. They can be installed like pavers (Snodgrass, 2006).

RESULTS AND DISCUSSIONS

The most influential parameter on vegetation layer is the leaf area index (LAI) that depends basically on the foliage density, the foliage geometric characteristics and on the plant height. Thus, the most important contribution of the vegetation layer to the thermal behaviour is shadow effect, both by the interception of solar radiation and reduction of roof surface temperatures (Theodosiou, 2003). Vegetation cover (LAI) and consequently the ability to produce shade can become reduced in certain periods of the life of the extensive green roof, for example during the plant growth period, which can last up to two years, or in water shortage periods, disease, or even because the type of plants, etc. Also it is known that because the type of plants and the low maintenance levels, extensive green roofs hardly reach 100% of the vegetation cover.

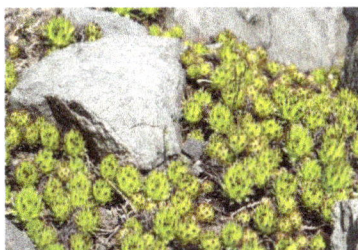

Figure 1: Sempervivium

Hardy succulents are the workhorses of extensive roofs and the primary plants for systems using a medium of 10, or less centimetres. Plants are native from dry locations, semi-dry locations, stony surfaces such as alpine environment. These kinds of plants have typical mechanisms to survive extreme conditions. Mechanisms like water storage organs, thick leaves, thick leaves surfaces, narrow leaves etc. They have unsurpassed ability to survive drought and wind conditions, store water in their leaves for extended periods and conserve water through a unique metabolic process. Hardly succulents like Sempervivum Fig. 1, Sedum, Talinum, Jovibarba, Delosperma are the only choices for thin substrate, non-irrigated, extensive green gardens with the greatest survivability (Snodgrass, 2006).

Figure 2: Phacelia Campanularia

Annuals should not be the dominant plant selection for extensive green roof, because they do not offer longevity required to make a project cost effective. They can be used as seasonal accents. They are required in places with regular rainfall, or in places with irrigation system. Annuals like Phacelia Campanularia Fig. 2, Portucala, Townsedia Eximia may be used on extensive green roofs as filters to provide quick colour during first grown season (Snodgrass, 2006).

Figure 3: Dianthus Dianthus Prairie Pink

Herbaceous perennials are the most desired plants for aesthetic reasons. They offer great colors, textures and season variability. On the other hand, they require deeper substrate and moisture than are found on most extensive green roofs. Some of them work very well on extensive roof installations. Dianthus Fig. 3, Phlox, Campanula Garganica, Teucreum, Allium, Potentilla, Achillea, Prunella, Viola, Origanum and some other low growing and shallow rooted perennials can be used, however, medium depth must be greater than 10 cm and has to have adequate water source. Few herbaceous perennials are evergreen, so if winter interest is a major design consideration for a roof, alternative must be provided so the brown vegetation is not so visible during its dormant period (Snodgrass, 2006).

Figure 4: Carex

Figure 5: Salvia

Short grasses are mostly popular for traditional intense green roofs and landscaping, but they still have their place in extensive green roofs. They do not have colorful bloomers like annuals and perennials, but they do have a lot of offer. They are more vertical, they add a texture and motion into a picture of roof, offer birds, insects habitat. They do need deeper depth of soil to accommodate their root system, some of them may have a Durant period during summer what means having a brown spot on the roof some time. Short grasses like Carex Fig.4, Festuca, Deschampsia are appropriate for extensive roofs.

Herbs as Thymus, Origanum, Salvia Fig. 5 and Allium have mostly their selected use on green roofs because of depth of soil that has to be more than 10 centimetres. On the other hand, their use is very specific because of their aroma, what is commonly used on buildings like restaurants, hospitals, residences, institutional buildings, they can be harvested for culinary, aromatic, therapeutic or educational purposes (Snodgrass, 2006).

CONCLUSIONS

Groundcovers like hardly succulents should be predominant plants used on extensive green roofs with a limited amount of accent plants. Groundcovers provide a rapid, reliable and cost-effective spread over the roof.

Accent plants like annuals, perennials, herbs and grasses while spectacularly during bloom, may not live more than five years on the roof.

In addition, they do not spread as rapidly as groundcovers, more plants are required to cover an area, but they offer seasonal interest. They may be replenished by periodic re-sowing.

Combination of these types of plants is the best way to come to this state. These combinations offer solution to make roof green all year long but with some colourful spots according to designed and used plants.

ACKNOWLEDGEMENTS

This work was supported by: VEGA 1/0202/15 Bezpečné a udržateľné hospodárenie s vodou v budovách tretieho milénia/ Sustainable and Safe Water Management in Buildings of the 3rd. Millennium.

REFERENCES

Dunett, N., 2004. Kingsbury, N., Planting green roofs & living walls, Timber press, Portland*Oregon, 9-82

Nakano, M., Rousseau, N., Henderson, D., 2014. Green roof plants: Establishment, viability and maintenance. http://www.kpu.ca/sites/default/files/Facilities%20Se rvices/Sustainability%20at%20KPU%20%20where% 20are%20we%20now%20%20Saima%20Zaida%20F inal%202014%2001%2030_0.pdf (20-02-2014)

Snodgrass, E. C., 2006. Green roof plants. Timber press, Portland*London, 11-88

Theodosiou, T., 2003. Summer period analysis of the performance of a planted roof as a passive cooling technique, Energy and Building

Thuring, C. E., 2014. Green roof plant response to different media depth under various drought conditions. http://horttech.ashspublications.org/content/20/2/395. full (20-02-2014)

22

INSIGHTS INTO MICROGREENS PHYSIOLOGY

Elena DELIAN, Adrian CHIRA, Liliana BĂDULESCU, Lenuța CHIRA

University of Agronomic Sciences and Veterinary Medicine of Bucharest, 59 Mărăşti Blvd, District 1, 011464, Bucharest, Rom ania
Corresponding author email: delianelena@yahoo.com

Abstract

In recent years people have a substantial interest for the consumption of fruits and vegetables characterized by a high content of bioactive substances. It is known that these are beneficial not only because they provides the necessary nutrients for human body, but, also have important effects on health. From this point of view, microgreens represent a new class of vegetables that can be considered as "functional foods". Although they have a short life cycle, one of the possibilities of manipulation of the bioactive compounds biosynthesis is to know that species physiology during germination, during the growth, as well as during postharvest. This review presents a general overview on some technological measures that may influence microgreens physiology, based on few recent research works carried out on these topics. There are reviewed data on: 1. beneficial effects of pre-sowing treatments; 2. the lights effects on microgreens physiology (in terms of quantity, but mostly quality of light) concerning the growth process, as well as accumulation of bioactive compounds; 3. measures to influence microgreens post-harvest physiology, to avoid the incidence of some microorganisms, to extend shelf life and to maintain their nutritional quality. Despite microgreens short life cycle, technological measures applied based on species physiology understanding are undoubtedly intended to result in increased the productivity, to obtain healthy products, with lower prices.

Key words: microgreens, physiology, bioactive compounds.

INTRODUCTION

In recent years people have a substantial interest for the consumption of fruits and vegetables characterized by a high content of bioactive substances. It is known that these are beneficial not only because they provides the necessary nutrients for human body, but, also have important effects on health (Beceanu, 2008; Galaverna et al., 2008; Mahima et al., 2013).

Microgreens are a new class of edible vegetables (Pinto et al., 2015), a very specific type which includes seedlings of edible vegetables, herbs or other plants, ranging in size from five to ten centimeters long (including stem and cotyledons) (Xiao et al., 2012).

Over the past few years, microgreens have gained popularity as a new culinary trend, being served as an edible garnish to embellish a wide variety of others dishes or a new salad ingredient (Frank and Richardson, 2009; Hedges and Lister, 2009; Chandra et al., 2012; Xiao et al., 2012; Kou et al., 2013; Pinto et al., 2015) thanks to their greater nutritional benefits (Chandra et al., 2012; Pinto et al., 2015).

Today, more attention is given to a healthy nutrition to prevent certain diseases (Márton et al., 2010). Microgreens are considered as "functional foods" which are food products that possess particular health promoting or disease preventing properties, that are additional to their normal nutritional values (Xiao et al., 2012). Nowadays, demand for these products is growing rapidly (Janovská et al., 2010; Samuoliené et al., 2012) and consumption is growing given their particular characteristics: unique color, rich flavor and appreciable content of bioactive substances (Brazaitytė et al., 2013; Kou et al., 2014; Sun et al., 2014; Brazaitytė et al., 2015a). These are also classified as a good source of minerals in the human diet (Pinto et al., 2015). However, it should be noted that due to the high price and high perishability, microgreens are not currently available to commercial terms in chain stores.

This crop has a quick production cycle (two to three weeks) and occupies very little space in greenhouse production (Kopsell et al., 2012;

Viršilė and Sirtautas, 2013). However, shorten the production cycle and thus reduce greenhouse production cost is one of the major goal of the current researches (Murphy and Pill, 2010). Another issue addressed was that the impact of the biotic stress agents on sprouts and microgreens, through the necessity of obtaining disease-free products (Pill et al., 2011; Xiao et al., 2014). The need to extend their shelf life is also a recent concern to researchers (Berba and Uchanski, 2012; Sasuga, 2014). Preharvest (Kou et al., 2014) or post-harvest (Lee et al., 2009) treatments can be effective means to achieve this objective (Kou et al., 2014).

The purpose of this review is to make a brief insight into microgreens physiology based on the recent research works carried out in this field. There are reviewed data on : 1. pre - sowing applied treatments as beneficial effects for seeds germination physiology; 2. lights effects on microgreens physiology (in terms of light quantity, but mostly light quality) referring to growth process, as well as accumulation of bioactive compounds; 3. measures to influence post-harvest physiology, to avoid the incidence of some microorganisms, to extend shelf life and to maintain microgreens quality.

MEASURES TO INCREASE GREENHOUSE PRODUCTION AND REDUCE COST PRICE

Growing microgreens is a relatively simple process, which does not require much time, energy and experience (Franks and Richardson, 2009). However, attention should be given to each step and besides there are necessary research to improve this type of culture, in terms of productivity, nutritional quality and last but not least the cost of production. Even if microgreens life cycle is very short, certain applied measures for improving seeds germination (as germination faculty, as well as germination velocity) are welcome, to determine a more rapid stabilization and to promote vigorous seedlings. To shorten the greenhouse production and thus the production cost, Murphy and Pill (2010) conducted a series of experiments on microgreen arugula (*Eruca vesicaria* subsp. *sativa*) grown in peat -lite (a soilless medium). Regarding pre-germination

treatments results have revealed that seed incubation in exfoliated vermiculite (1.12 g seed in 157 g vermiculite) moistened with 2 g H_2O g^{-1} dry weight vermiculite for 1 day at 20 ^{0} C resulted in a 21% increase in shoot fresh weight by 14 days after planting, as against untreated seed before sowing. Pre-germinated seed germination was 81.5% and at the time of sowing the radicle already had an average length of 2 mm. In terms of sowing density it has proven that high density caused an increase in shoot fresh weight m $^{-2}$ at 10 days after planting, compared to using a lower density. Regarding fertilization, the experiments showed that the most economical measures that have induced an increase in fresh weight m^{-2} were those based on daily use of 150 mg N L $^{-1}$ or daily solution fertilisation that contained 75 mg N L $^{-1}$ plus a pre-plant media incorporation of 1,000 mg N L $^{-1}$ from $Ca(NO_3)$;

Impact of seed treatment was experienced as well for others species: radish (*Raphanus sativus*), kale (*Brassica napus* var. *pabularia*), and amaranth (*Amaranthus tricolor*) (Lee and Pill, 2005). Pretreatment for 3 days in vermiculite (at 12 ^{0} C and -1.0 MPa), removing from the vermiculite and then dried the seeds to the initial weight (before sowing) although not greatly influenced germination faculty, had a positive impact on germination velocity at all studied species. The emergence of radish and kale was faster than untreated seeds although 13 days after planting there has not been an increase of shoot dry weight. In contrast, the amaranth shoot dry weight was increased. If seeds have been exposed to a two-step primed treatment: 3 days at 50% H_2O and then germination 1 day in vermiculite at 150 % H_2O, the shoot dry weights increased by 20, 49, and 84% for radish, kale, and amaranth, respectively, as against to those from non-treated seeds.

Also, Lee et al. (2004) examined some possibilities to advance the establishment of table beet or chard (*Beta vulgaris* L.) for greenhouse microgreens production. Germination of seeds in fine grade vermiculite and sowing the germinated seeds with vermiculite mixture caused a more rapid seedlings emergence. If before this process seeds were primed and soaked with hydrogen peroxide, there has been no progress in terms of emergence advancement or growth. On the other hand, germinating the seeds in shallow (4 cm deep) vermiculite (150% initial water, 1 seed : 3 vermiculite dry weight ratio, 27 ^{0}C) for 2 days (table beet) or 3 days (chard) resulted in 0.33-fold and 2.79-fold greater shoot fresh weight, respectively, at 11 days after planting than was achieved by sowing untreated seeds.

LIGHT INFLUENCE ON MICROGREENS PHYSIOLOGY AND THEIR NUTRACEUTICAL QUALITY

One of the possibilities to manipulate the physiology of the plant, including bioactive compounds is controlling environmental conditions (Murchie et al., 2011; Jones, 2014). Light is one of the main external factors absolutely necessary for the photosynthetic organisms, as it is a source of energy and information from the environment (Murchie and Niyogi, 2011; Fortunato et al., 2015). All oxygenic photosynthetic organisms need strategies for maintaining the balance between efficient light harvesting, photochemistry and photoprotection from excess light (Goss and Lepetit, 2015; Quaas et al., 2015). Light intensity and its quality not only influence the rate of photosynthesis in plants, but also the accumulation of different organic compounds (in terms of their quantity and quality), including production of secondary plant compounds (Murchie et al., 2011; Brazaitytė et al. 2015 a,b).

Vegetables are designated as healthy foods of the millennium or nutraceutical foods of the century (for reviews see Rahal et al., 2014). Microgreens contain higher concentrations of bioactive compounds such as vitamins, minerals, and antioxidants, than mature greens (Janovská et al., 2010; Xiao et al., 2012).

According to Kopsell and Kopsell (2006) one important class of phytochemicals is the carotenoids. Recent studies of Brazaitytė et al. (2015) from the view point of light intensity and its quality influence emphasized that *Brassicaceae* microgreens accumulated more total carotenoids in the case of 330–440 µmol m^{-2} s^{-1} wavelengths levels, as against to 220 µmol m^{-2} s^{-1} , considered as a normal one.

On the other hand, changes of carotenoids content can be achieved by changing of the light spectral composition, relative to the species. Thus, in mustard, supplementation with green light determines an increase in lutein / zeaxanthin and beta- carotene. In the case of other species (red pak choi and tatsoi), the standard blue, red, and far- red light are favourable.

As stated Brazaitytė et al. (2015b), additional application of UV-irradiation for basal lighting with light emitting diodes may lead to an improvement in antioxidant characteristics at microgreens depending on the species. A significant beneficial effect was induced in the case of wavelengths of 366 nm and 390 nm at a photon flux density of 12.4 µmol m^{-2} s^{-1}.

Short-term pre-harvest supplemental 638-red lighting on *Perilla frutescens* (L.) Britton (an annual herbaceous edible and medical plant of *Lamiaceae* Lindl. family naturally growing in the East Asia) on the one hand may reduce the quality of plants, due to the increase in major antioxidants content (e.g. anthocyanins and acid ascorbic) and on the other hand the decrease of nitrate (Brazaitytė et al., 2013). The authors did not show any effect of treatment on DPPH (1,1-diphenyl-2-picrylhydrazyl) free radical scavenging activity and flavonols index, while α-tocopherol content decreased.

Sirtautas and Samuolienė (2013) noticed that the red light-emitting diode (LED) and 24 h photoperiod effect for nitrate and antioxidant contents red baby leaf lettuce is variety – dependant and proper lighting strategies should be selected seeking to cultivate lettuce with optimal contents of phytochemical compounds. However, light spectra and photoperiod are suitable tools seeking to create mild photo stress for plants, with the aim to enhance the contents of antioxidant phytochemicals. In the case of borage (*Borago officinalis* L.) Viršilė and Sirtautas (2013) showed that 440 µmol m^{-2}s^{-1} LED illumination is a recommendable light intensity for microgreens production with optimal growth and nutrient contents. Lower irradiance levels result in significant accumulation of nitrates, decreased fresh weight and elongated hypocotyls, when the highest investigated 545 µmol m^{-2} s^{-1} photosynthetic photon flux density level exceeds plant tolerance and is associated with the decreased growth parameters and slightly lower antioxidant phytochemical contents.

As regard as the impact of supplementary short-term red LEDs lighting on the antioxidant properties of microgreens, Samuolienė et al. (2012) established that natural antioxidant compounds varied in function of the studied species. Thus, the gradual decline of this activity was registered, from a maximum value determined at pea (then lower values for broccoli, borage, mustard, amaranth, basil, kale, beet, parsley, in the order presented here), to very low values for tatsoi. Differentiation was recorded in terms of content in phenols due to additional red light assuring, so variable content was noticed, with an increase (from 9.1% to 40.8% in the tatsoi mustard) or a decrease (with 14.8% at amaranth) .

An obvious variability was also noted in terms of ascorbic acid and anthocyanins content. For example, ascorbic acid in amarantus increased by 79.5%, while in basil it decreased by 53.9%. The amount of anthocyanins increased significantly in broccoli (45.1%), decreased markedly in borage (51.8%), while in basil it was not significantly influenced.

For *Brassicaceae*, Samuolienė et al. (2013) noticed that, the best conditions for growth and nutritional quality (higher leaves, higher content of anthocyanins, phenolics, antioxidant activity, lower nitrate content) were at 320-440 μmol cm^{-2} s^{-1}. In the case of 545 μmol cm^{-2}s^{-1} there has not been a significant positive impact on these indicators.

Antioxidant activity was found both in common and tartary buckwheat microgreens. High levels of flavonoids, carotenoids, and α-tocopherol were detected as well. Higher amount of flavonoids was detected in tartary buckwheat microgreens. No significant differences were detected between common and tartary buckwheat microgreens in content of phenolic acids. Microgreens of both common and tartary buckwheat represent potential nutritional sources of alternative vegetable in the Czech Republic (Janovská et al., 2010). All the vegetables (microgreens, sprouts and leafy greens) of both varieties of buckwheat can be regarded as a potent source of phenolics (rutin, quercetin, vitexin, isovitexin, orientin isoorientin and chlorogenic acids) and has high antioxidant activities.

Nepalese strain buckwheat vegetables contain high phenolics with higher biological (antioxidant and α-glucosidase inhibition) activity and can be used as an alternative food. Therefore, mass production of more and more buckwheat food products should be encouraged and included in the daily diet, which would help the people to prevent diabetes and many other diseases caused by the free radicals (Sharma et al., 2012).

That microgreens are suitable sources for bioactive substances was also demonstrated by Sun et al and (2013) from studies conducted on five species of *Brassica* vegetables. Considering the use of advanced techniques, such as ultrahigh-performance liquid chromatography photodiode array multistage high-resolution mass spectrometry it was possible to identify 164 polyphenols: anthocyanins (30), flavonol glycosides (105) and hydroxycinnamic and hydroxybenzoic acid derivatives (29). It was also revealed that the phenolic profile was more complex and also the types of substances present in microgreens were varied unlike mature plants.

The lack of scientific data about the nutritional content of microgreens prompted Xiao et al. (2012) to undertake extensive research on concentrations of ascorbic acid, carotenoids, phylloquinone, and tocopherols in 25 commercially available microgreens. Red cabbage, cilantro, garnet amaranth, and green daikon radish had the highest concentrations of ascorbic acids, carotenoids, phylloquinone, and tocopherols, respectively. In comparison with nutritional concentrations in mature leaves (USDA National Nutrient Database), the microgreen cotyledon leaves possessed higher nutritional densities.

Increased nutritional value of microgreens can be also achieved by light management as demonstrated results of Kopsell et al. (2012). The authors' conclusion was that simple applications of short duration high light resulted in biochemical shifts in xanthophyll cycle pigment concentrations in the microgreens, most notably increases in zeaxanthin (characterized by antioxidant and protective effects of vision) may have beneficial effects for the human diet. Thus, the xanthophyll cycle pigments (zeaxanthin + antheraxanthin + violaxanthin) (known in plants as having a major role in the dissipation of excess light absorbed energy) by mustard (*Brassica juncea* L ' Florida Broadleaf ') microgreens exposure to increased light intensity before tissue harvest resulted in increased concentration in zeaxanthin levels. Treatment with 463 μmol photons $m^{-2} s^{-1}$ after the appearance of the first true leaves (light treatment had accumulated 36 h during the photoperiod) resulted in a significant decrease in chlorophyll (a and b) content, β - carotene, and neoxanthin, which indicated the incidence of stress. Lutein content remained unchanged, while the concentration of zeaxanthin and antheraxanthin increased.

Also, management of LED lighting technology through preharvest, short-duration blue light acted to increase important phytochemical compounds influencing the nutritional value of broccoli microgreens (Kopsell and Sams, 2013). At 13 days after sowing broccoli plantlets were treated for 5 days before being harvested using: 1) red and blue LED light (350 μmol $m^{-2} s^{-1}$); or 2) blue LED light (41 μmol $m^{-2} s^{-1}$). Determinations done on biological material highlighted that short-term treatment before harvest with blue light significantly increased shoot tissue β-carotene, violaxanthin, total xanthophyll cycle pigments, glucoraphanin, epiprogoitrin, aliphatic glucosinolates, essential micronutrients of copper, iron, boron, manganese, molybdenum, sodium, zinc, and the essential macronutrients of calcium, phosphorus, potassium, magnesium, and sulphur.

The nutritional value of microgreens was also demonstrated by Pinto et al. (2015), who carried out a comparative study as regard as microgreens and mature lettuces mineral profile. Studies have led to excellent results, not only in scientific terms but rather in terms of practical utility. Microgreens had higher content of Ca, Mg, Fe, Mn, Zn, Se and Mo than mature lettuces, although the former possessed higher N, P and K content. Therefore, microgreens may be considered a good source of minerals. In addition, very low content of NO_3 justifies their recommendation as being safety in the human diet, especially for children, in order to complete their requirements for minerals.

HEALTHY MICROGREENS WITH A LONGER SHELF LIFE

Microgreens senesce rapidly after harvest and have typically a very short shelf-life (1-2 days) at ambient temperature, due to the sudden disruption of plant growth at a very early stage (Guo and Gan, 2012; Xiao et al., 2014).

At broccoli microgreens (Sun et al., 2015), obtaining a high production and extend shelf life has been demonstrated to be possible by applying preharvest treatment with 10 mM calcium chloride. Metabolome analysis obtained by ultra-high-performance liquid chromatography with mass spectrometry led to the conclusion that glucosinolates were the main group of chemical compounds in the treatment variant and the amount of aliphatic and indolic glucosinolates was significantly higher compared with control.

Extending shelf life (by optimize postharvest handling conditions) is a permanent goal of all those who produce and sell horticultural products of any kind. There are concerns on this topic (Berba and Uchanschi, 2012: Chandra et al., 2012; Kou et al., 2013; Kou et al., 2014; Xiao et al., 2014). It should be noted that, recently Sasuga (2014) patented a method for providing a microgreens product with significantly longer shelf life than most other microgreens products, is claimed, where the product microgreens product has a shelf life of at least 10 days.

Researches carried out by Xiao et al. (2014) using radish seeds inoculated with *Escherichia coli* O157: H7 or O104: H4 and *E. coli* populations on harvested products (sprouts and microgreens) highlighted the followings: the proliferation of the bacteria was done in the case of both products, but on microgreens bacteria was significantly reduced. During sprouting, *E. coli* O157: H7 and O104: H4 has reached levels of 5.8-8.1 log cfu / g and 5.2-73 log cfu / g, respectively, depending on the inoculation level (1.5-4.6 log cfu / g and 0.8-43 log cfu / g on radish seeds, respectively), while for microgreens values ranged from 0.8 to 4.5 log cfu / g and from 0.6 to 4.0 log cfu / g, respectively. At table beet (*Beta vulgaris* L.) (,of 'Early Wonder Tall Top') in order to biological control of the fungus *Pythium aphanidermatum* (Edson) Fitzp., Pill et al.

(2011) experienced the influence of the seeds treatment by using equal amounts of antagonistic fungus *Trichoderma harzianum* Rifai strain KRL-AG2 G41 and *T. virens* strain G-41 (*ThTv*) at 0 , 0.25 , 0.50 , 0.75 or 1.00 mg per seed ball. Four days before planting, the peat-lite was inoculated with *P. aphanidermatum* at 0, 0.5 and 1.0x the rate that resulted in 96% damping-off when non-*T. harzianum* -*T. virens*-treated dry seed balls were sown in peat-lite containing 1.0 *P. aphanidermatum*. Decreasing incidence of damping-off with increasing *T.hartzianum-T.virens* application to seed balls or growth media was associated with increasing shoot fresh weight m^{-2} at 14 days after planting, a response attributable to increased percentage plant survival and not to a *ThTv* growth-promoting effect.

Post-harvest tah tasai Chinese cabbage young leaf vegetable were washed in cold (5° C) and warm (25° C) chlorinated water (0, 50 or 100 mg L^{-1} free chlorine for 90 sec), then packaged in polypropylene film bag and stored for 8 days at 15° C. Chlorinated water treatment at 5° C had a more beneficial effect on visual quality, weight loss, SPAD (greenness) value change than 25° C chlorinated water treatment. No significant difference was found in aerobic plate count on the surface of microgreens after 3-day storage period. Chlorinated water either at 5° C or 25° C with 50-100 mg L^{-1} free chlorine significantly reduced aerobic plate count during the initial period of storage (up to 2 days) (Lee et al., 2009).

As Hodges and Toivonen (2008) noticed, the two most important storage parameters for postharvest shelf life are storage temperature and atmospheric composition. For example, Kou et al. (2013) indicated that for buckwheat microgreens storage temperature significantly affected the changes in O_2 and CO_2 composition, tissue electrolyte leakage and microbial growth during storage. The authors suggested that buckwheat microgreens should be stored at 5° C with moderately high O_2 (14.0-16.5 kPa) and moderately low CO_2 (1.0-1.5 kPa) content to maintain optimal quality and maximal shelf life. Another useful measure to extend the microgreens shelf life is controlling respiration rates (Berba and Uchanschi, 2012). Post-harvest experiments on

arugula (*Eruca sativa*), radish (*Raphanus sativus*), and red cabbage (*Brassica oleracea* var. *rubra*) designed to lower respiration rates, increase shelf life and will allow for transportation to larger markets.

In order to increase biomass and delays senescence of broccoli microgreens, Kou et al. (2014) studied the effect of pre-harvest daily sprays (for 10 days) with calcium chloride solution (1, 10 and 20 mM) as compared to the control (water). After harvesting fresh-cut microgreens packaged in sealed polyethylene film bags were stored on headspace atmospheric conditions. Visual quality and tissue membrane integrity were evaluated on days 0, 7, 14, and 21 during 5 degrees C storage. It was found that treatment with 10 mM calcium chloride resulted in the increase of biomass by more than 50% and the calcium content was triplet, as against the control. Also, this treatment led to an increase in superoxide dismutase and peroxidase enzyme activity, decreased electrolyte leakage in the plant tissue, improved visual quality and reduced microbial growth during storage. Following research in Tan Tasai, Chinese cabbage (*Brassica campestris* var. *marinosa*) Chandra et al. (2012) reported that citric acid and ascorbic acid mixed solution (0.25 % w/v) in addition with ethanol spray (50%) can be used to replace chlorine used for washing microgreens and polyethylene film can provide more benefits for packaging microgreens.

Storage temperature, packaging film, and wash treatment were investigated by Xiao et al (2014) at daikon radish (*Raphanus sativus* L. var. *longipinnatus*) microgreens. Accordingly, studies conducted in dynamics during storage highlighted that storage temperature significantly affected package atmosphere, product quality and shelf life. The optimal temperature for storage of radish microgreens with no chilling injury was proved to be 1^{0}C. Film oxygen transmission rate significantly affected O_2 and CO_2 composition, but this did not significantly affect quality attributes during 28 days of storage at 1 ^{0}C. Chlorine wash treatment (100 mg L^{-1}) significantly reduced initial microbial populations.

As Artées-Hernández (2013) recently noticed, attention should be focused on "eco-innovative emerging alternative" to prolong the shelf life without losing quality characteristics specific to the fresh product.

CONCLUSIONS

Microgreens represent a new class of vegetables that can be considered as "functional foods". The manipulation of bioactive compounds biosynthesis and preserve nutritional quality after harvesting may be based on recent research works carried out on these topics so far, but mainly those that come as a necessity. The facts reviewed here indicate that there are promising results in terms of: 1. beneficial effects of pre-sowing treatments; 2. the lights effects on microgreens physiology (in terms of quantity, but mostly quality of light) concerning the growth process, as well as accumulation of bioactive compounds; 3. measures to influence microgreens post-harvest physiology, to avoid the incidence of some microorganisms, to extend shelf life and to maintain their nutritional quality.

Despite microgreens short life cycle, technological measures applied based on species physiology understanding are undoubtedly intended to result in increased the productivity, to obtain healthy products, with lower prices.

Undoubtedly, further studies are needed in terms of content in phytonutrients. These data may provide a scientific basis for evaluating nutritional values of microgreens and contribute to food composition database. Moreover, such results may be used as a reference for health agencies' recommendations and consumers' choices of fresh vegetables (Xiao et al., 2012).

REFERENCES

Artés-Hernández F., Gómez P.A., Artés F. 2013. Unit processing operations in the fresh-cut horticultural product industry: quality and safety preservation, in Food Quality, Safety and Technology, Ed. Lima, G.P.P. and Vianello, F., Springer, 35-53.

Beceanu D. 2008. Nutritive, nutraceutical, medicinal and energetic value of fruits and vegetables. Cercetări Agronomice în Moldova, Vol. XLI, 4 (136), 65-81.

Berba K.J., Uchanski M.E. 2012. Post-harvest physiology of microgreens. Journal of Young Investigators, Vol.24, 1-5.

Brazaityté A., Jankauskiené J., Novičkovas A. 2013. The effects of supplementary short –term red LEDs

lighting on nutritional quality of Perilla frutescens L. microgreens. Rural Development, 54-57.

Brazaityté A., Sakalauskiené S., Samuoliené G., Jankauskiené J., Viršile A., Novičkovas A., Sirtautas, R., Miliauskiené J., Vaštakaité V., Dabašinskas L., Duchovskis P. 2015a. The effects of LED illumination spectra and intensity on carotenoid content in Brassicaceae microgreens. Food Chemistry, Vol.173, 600-606.

Brazaityté A., Viršile A., Jankauskiené J., Sakalauskiené S., Samuoliené G., Sirtautas, R., Novičkovas A., Dabašinskas L., Miliauskiené J., Vaštakaité V., Bogdonovičiené A. ,Duchovskis P. 2015b. Effect of suplimental UV-A irradiation in solid-state lighting on the growth and phytochemical content of microgreens. Int. Agrophys. Vol.29, 13-22.

Chandra D., Kim J.G., Kim Y.P. 2012. Changes in microbial population and quality of microgreen treated with different sanitizers and packaging films. Hort. Environ. Biotechnol., Vol. 53, 32-40.

Fortunato A.E., Annunziata R., Jaubert M., Bouly J.P., Falciatore A. 2015. Dealing with light: The widespread and multitasking cryptochrome/photolyase family in photosynthetic organisms. Journal of Plant Physiology, Vol. 172, 42-54.

Franks E., Richardson J. 2009. Microgreens. A guide to growing nutrient-packed greens. Published by Gibbs Smith, Layton, Utah.

Galaverna G., Di Silvestro G., Cassano A., Sforza S., Doceana A., Drioli E., Marchelli R. 2008. A new integrated membrane process for the production of concentrated blood orange juice: effect on bioactive compounds and antioxidant activity. Food Chem. Vol. 106,1021–30.

Goss R., Lepetit B. 2015. Biodiversity of NPQ. Journal of Plant Physiology, Vol. 172, 13-32.

Guo Y.F., Gan S.S. 2012. Convergence and divergence in gene expression profiles induced by leaf senescence and 27 senescence-promoting hormonal, pathological and environmental stress treatments. Plant Cell Environ., Vol. 35, 644–655.

Hodges D. M., Toivonen P. M. A. 2008. Quality of fresh-cut fruits and vegetables as affected by exposure to abiotic stress. Postharvest Biology and Technology, Vol. 48, 155-162.

Hedges L.J., Lister C.E. 2009. Nutritional attributes of some exotic and lesser known vegetables. Plant and Food Research Confidential Report No. 2325.

Janovská D., Štočková L., Stehno Z. 2010. Evaluation of buckwheat sprouts as microgreens. Acta Agriculturae Slovenica, 95–2, 157 - 162.

Jones H.G. 2014. Plant and microclimate a quantitative approach to environmental physiology. Third Edition. Cambridge University Press.

Kopsell D. A., Kopsell D. E. 2006. Accumulation and bioavailability of dietary carotenoids in vegetable crops. Trends in Plant Science, Vol.11(10), 499–507.

Kopsell D.A., Pantanizopoulos N.I., Sams C.E., Kopsell D.E. 2012. Shoot tissue pigment levels increase in 'Florida Broadleaf' mustard (Brassica juncea L.) microgreens following high light treatment. Scientia Horticulturae 140, 96–99.

Kopsell D.A., Sams C.E. 2013. Increases in shoot tissue pigments, glucosinolates, and mineral elements in sprouting broccoli after exposure to short-duration blue light from light emitting diodes. Journal of the American Society for Horticultural Sciences, Vol.138, 31-37.

Kou L., Luo Y., Yang T., Xiao Z., Turner E.R., Lester G.E., Wang Q., Camp M.J. 2013. Postharvest biology, quality and shelf life of buckwheat microgreens. Food Science and Technology, Vol. 51, 73-78.

Kou L.P., Yang T.B., Luo Y.G., Liu X.J., Huang L.H., Codling E. 2014. Pre-harvest calcium application increases biomass and delays senescence of broccoli microgreens. Postharvest Biology and Technology, Vol. 87, 70-78.

Lee J.S., Pil, W.G., Cobb B.B., Olszewski M. 2004. Seed treatments to advance greenhouse establishment of beet and chard microgreens. Journal of Horticultural Science and Biotechnology, Vol. 79, 565-570.

Lee J.S., Pill W.G. 2005. Advancing greenhouse establishment of radish, kale and amaranth microgreens through seed treatments. Horticulture, Environmentand Biotechnology, Vol. 46, 363-368.

Lee J.S., Kim J.G., Park S. 2009. Effects of chlorine wash on the quality and microbialpPopulation of 'Tah Tasai' chinese cabbage (Brassica campestris var. narinosa) microgreen. Korean Journal of Horticultural Science and Technology, Vol. 27, 625-630.

Mahima, Amit Kumar Verma, Ruchi Tiwari, K. Karthik, Sandip Chakraborty, Rajib Deb and Kuldeep Dhama. 2013. Nutraceuticals from fruits and vegetables at a glance: A Review. Journal of Biological Sciences, Vol. 13, 38-47.

Márton M., Mándoki Zs., Csapó J. 2010. Evaluation of biological value of sprouts. Fat content, fatty acid composition. Acta Univ. Sapientiae Alimentaria, Vol. 3, 53-65.

Murchie E.H., Niyogi K.K. 2011. Manipulation of photoprotection to improve plant photosynthesis. Plant Physiology, Vol.155, 86-52.

Murphy C.J., Pill W.G. 2010. Cultural practices to speed the growth of microgreen arugula (roquette; Eruca vesicaria subsp sativa). Journal of Horticultural Science and Biotechnology, Vol. 85, 171-176.

Pill W.G., Collins C.M., Gregory N., Evans T.A. 2011. Application method and rate of Trichoderma species as a biological control against Pythium aphanidermatum (Edson) Fitzp. in the production of microgreen table beets (Beta vulgaris L.). Scientia Horticulturae, Vol.129, 914-918.

Pinto E., Almeida A.A., Aguir A.A., Ferreira I.M.P.L.V.O. 2015. Comparison between the mineral profile and nitrate content of microgreens and mature lettuces. Journal of Food Composition and Analysis, Vol. 37, 38–43.

Quass T., Berteotti S., Ballottari M., Flieger K., Bassi R., Wilhelm C., Goss R. 2015. Non-photochemical quenching and xanthophyll cycle activities in six green algal species suggest mechanistic differences in the process of excess energy dissipation. Journal of Plant Physiology, Vol. 172, 92-103.

Rahal A., Mahima A.K., Verma R., Kumar A., Tiwari R.M, Kapoor S., Chakraborty S., Dhama K. 2014. Phytonutrients and nutraceuticals in vegetables and their multi-dimensional medicinal and health benefits for humans and their companion animals: A Review. Journal of Biological Sciences, Vol. 14, 1-19.

Samuolienė G., Brazaitytė A., Sirtautas R., Sakalauskienė S., Jankauskienė J., Duchovskis P. 2012. The impact of supplementary short-term red LED lighting on the antioxidant properties of microgreens. Acta Hort. (ISHS) 956, 649-656.

Samuolienė G., Brazaitytė A., Jankauskienė J., Viršile A., Sirtautas R., Novičkovas A., Sakalauskienė S., Sakalauskaité J., Duchovskis P. 2013. LED irradiance level affects growth and nutritional quality of Brassica microgreens. Centr. Eur. J. Biol., Vol. 8, 1241- 1249.

Sasuga D.G.2014. Providing microgreens e.g. celery product with significantly longer shelf life than most other microgreens products. Patent Number(s): WO2014117034-A2; US2014212549-A1.

Sharma P., Ghimeray A.K., Gurung A., Jin C, W., Rho H.S., Cho, D.H. 2012. Phenolic contents, antioxidant and α-glucosidase inhibition properties of Nepalese strain buckwheat vegetables .African Journal of Biotechnology, Vol. 11(1),184-190.

Sirtautas R., Samuolienė G. 2013. The effect of red-LED lighting on the antioxidant properties and nitrates in red baby leaf lettuces. Rural Development, 237-240.

Sun J., Xiao Z., Lin L., Lester G.E., Wang Q., Harnly J.M., Chen P. 2013. Profiling polyphenols in five Brassica species microgreens by UHPLC-PDA-ESI/HRMS. J. Agric. Food. Chem., Vol. 61, 10960-10970.

Sun J., Kou L., Geng P., Huang H., Yang T., Luo Y., Chen P. 2015. Metabolomic assessment reveales an elevated level of glucosinolate content in CaCl2 treated broccoli microgreens. J. Agric. Food Chem., Vol. 63 (6), 1863–1868.

Viršilė A., Sirtautas R. 2013. Light irradiance level for optimal growth and nutrient contents in borage microgreens. Rural Development, 272-275.

Xiao Z.L., Lester G.E., Luo Y.G., Wang Q. 2012. Assessment of vitamin and carotenoid Concentrations of emerging food products: Edible microgreens. Journal of Agricultural and Food Chemistry, Vol. 60, 7644- 7651.

Xiao Z.L., Nou X.W., Luo Y.G., Wang Q. 2014. Comparison of the growth of Escherichia coli O157: H7 and O104: H4 during sprouting and microgreen production from contaminated radish seeds. Food Microbiology, Vol.44, 60-63.

Xiao Z. L., Luo Y., Lester G.E., Kou L., Yang T., Wang Q. 2014. Postharvest quality and shelf life of radish microgreens as impacted by storage temperature, packaging film, and chlorine wash treatment. Food Science and Technology, Vol. 55, 551-558.

23

DYNAMICS OF PHYSIOLOGICAL PROCESSES IN TOMATOES DURING THE PHENOLOGICAL STAGES

Aurelia DOBRESCU[1], Gheorghița HOZA[1], Daniela BĂLAN[1]

[1]University of Agronomic Sciences and Veterinary Medicine Bucharest 59 Marasti Blvd., 11464 Bucharest, Romania

Corresponding author email: balan.dana@gmail.com

Abstract

Tomato (Solanum lycopersicum L.), known to belong to the Solanaceae family, is considered one of the most important vegetable in the world since the fruits are widely consumed either fresh or processed. The ripe fruits are a valuable source of vitamin C, carotenoids and minerals such as iron and phosphorous that is daily required for a healthy diet. Fruit growth and ripening are the result of multiple physiological and metabolical processes that occur during the plant development. A thorough knowledge of the physiological characteristics of tomato plants is necessary to improve the technology of cultivation under greenhouse conditions. Research regarding the intensity of physiological processes as photosynthesis, respiration, transpiration have been made on some tomato hybrids cultivated in protected spaces in different phenological stages of plant development: in the vegetative growth period, at the flowering time and at the fruiting time. The measurements were performed with electronic analyzer LCA-4. It has been noticed that dynamics of the physiological processes varies depending on the hibrid type and on the development phenophase.

Key words: tomato, phenology, photosynthesis, respiration, transpiration.

INTRODUCTION

Tomato (*Solanum lycopersicum* L.), known to belong to the *Solanaceae* family, is considered one of the most important vegetable in the world since the fruits are widely consumed either fresh or processed. Beside the high nutritional value, the ripe tomato fruits are a valuable source of vitamin C, carotenoids and minerals such as iron and phosphorous that are daily required for a healthy diet (Nour et al., 2013).

Fruit growth and ripening are the result of multiple physiological and metabolical processes that occur during the plant development. Leaves are considered to be the main providers of carbon for fruit growth (Hetherington et al., 1998). Therefore, the major functions of the leaves was studied in order to relate the influence of different cultivation technologies or varying environmental conditions on fruit growth and development. A thorough knowledge of the physiological characteristics of tomato plants is necessary to improve the technology of cultivation. The physiological parameters depends on genetics, environmental factors (temperature, light, water and nutrient availability, air composition), agricultural techniques (Schwarz et al., 2002; Islam, 2011; Zhu et al., 2012).

Recently new tomato hybrids with improved nutritional content and potential health benefits are being developed. Consequently, it has become increasingly important to assess their physiological parameters in order to recommend the use of certain cultivation technologies. For this purpose study of the intensity of physiological processes as photosynthesis, respiration, transpiration was performed on some tomato hybrids cultivated in protected spaces (greenhouse). The determinations were made in different phenological stages of plant development (growth, flowering and fruiting phenophase) so that some peculiarities of selected tomato hybrids to be emphasized.

MATERIALS AND METHODS

Four tomato hybrids from collection of Faculty of Horticulture (USAMV Bucharest) were investigated: Principe Borghese, Maressa, Izmir and Ruxandra.

Principe Borghese hybrid is a cherry tomato with determined growth, small pear-shaped fruits, ideal for consumption in fresh and preserved condition.

Maressa hybrid is characterised by undetermined growth, uniform round fruits of middle size (150-170 g), suitable for consumption in a fresh state.

Izmir and Ruxandra are tomato hybrids with undetermined growth, which produce big round fruits (180-200 g).

The selected tomato hybrids were cultivated in protected systems (greenhouse), that provided controlled conditions for plant growth, so that the determinations were made at 720-880 μmols/m^2/s light intensity and a temperature of 22-24°C.

Photosynthesis, respiration and transpiration rates were determined with LCA-4 analyzing portable system (ADC Bioscientif, UK) on the fifth leaf from the top of plant. The measurements were made on 10 tomato plants randomly chosed in the greenhouse and average of these 10 measurements was calculated. The obtained results were expressed in μmols CO_2/m^2/s for photosynthesis and respiration rate and μmols H_2O/m^2/s for transpiration rate.

RESULTS AND DISCUSSIONS

The performed research approached the variations of some physiological parameters of certain tomato hybrids during the development stage, in different phenological phases: vegetative growth, flowering and fruiting phases.

The photosynthesis process

Tomatoes are included in C_3 photosynthetic type. The intensity of photosynthesis determines growth and development of plants, so directly influences the yield quantity and quality (Burzo and Dobrescu, 2005).

The obtained results (Figure 1) pointed out that in *the vegetative growth stage* it can be noticed that the most reduced photosynthetic rate was determined at Principe Borghese hybrid (3.21 μmols CO_2/m^2/s), while Izmir hybrid registered a 1.5 times higher value (4.88 μmols CO_2/m^2/s).

Also Maressa and Ruxandra hybrids reached similar values, no significant differences were reported by comparison with Izmir hybrid.

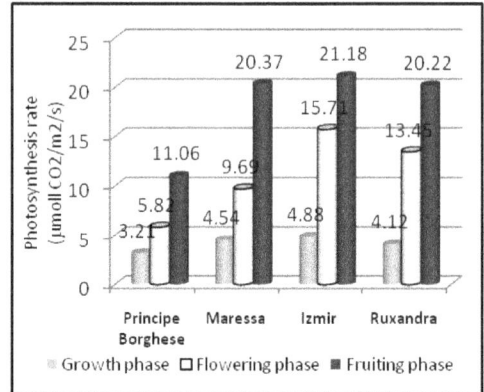

Figure 1. Dynamics of photosynthesis in the leaves of selected hybrids

The photosynthesis of tomato hybrids has been monitored also in *the flowering phenophase*, which debuted in May. In this period Izmir hybrid was noted with an increased value of photosynthesis rate (15.71 μmols CO_2/m^2/s), which was 1.17 times higher compared to Ruxandra hybrid and 2,7 times higher than the one registered by the Principe Borghese hybrid. Comparing the data obtained in the two analyzed phenophases, it can be appreciated that in the flowering stage the process of photosynthesis increased in all selected hybrids, but with a different rhythm: it was 3.21 times more intense at Izmir hybrid, 2.44 times higher at Maressa and only 1.8 times higher at Principe Borghese. This dynamics of photosynthesis can be correlated with the achievement of the growth of leaves, which reached the characteristic dimensions. These results are in according with Ludwig and Withers (1984), which determined the highest value of photosynthesis intensity when the tomato leaves reached 30-50% of their maximum area.

Also in *the fruiting phase* were made determinations of the photosynthetic rate, which reached higher values compared to flowering phase in all the selected hybrids because the formation and growth of the fruits stimulate the photosynthesis process. Thus, the photosynthesis increased by 2.1 times at Maressa hybrid, by 1.9 times at Principe Borghese hybrid and by 1.5 times at Ruxandra hybrid.

Izmir hybrid recorded the highest photosynthetic rate in this phenophase (21.18

μmols $CO_2/m^2/s$) compared to the other analyzed hybrids. However, in this hybrid the photosynthesis process was stimulated only 1.3 times by fructification, while the flowering determined a 3.3 times more intense rate of photosynthesis in comparison with the growth stage.

It is notable Izmir hybrid as having an elevated biological potential, given the high value of the photosynthetic intensity performed during monitored phenophases.

The transpiration process

Transpiration process consists in removal of the water excess by the plants, thus avoiding supersaturation of the cells with watter and overheating (thermoregulator role). Most importantly, transpiration generates the suction force of the leaf, which is involved in roots activity of water and minerals absorption. Thus it can be appreciated that transpiration and photosynthesis are related processes, as the suction force of leaf provides the raw materials needed to develop leaf photosynthesis (Burzo et al., 2004).

The obtained data (Figure 2) indicated that in *the growth phenophase* the transpiration rate varied between 2.33 μmols $H_2O/m^2/s$ in Principe Borghese hybrid and 3.31 μmols $H_2O/m^2/s$ in Izmir hybrid. Comparing the values determined for the selected hybrids, it was noticed that Izmir hybrid performed a transpiration rate by 1.42 times more intense than Principe Borghese, by 1.38 times higher than Maressa and by 1.4 times higher than Ruxandra hybrids.

In *the flowering phenophase* an increasing of transpiration rate can be observed compared to the growth phase in all the selected tomato hybrids: by 1.36 times at Ruxandra, by 1.11 times at Maressa, but insignificantly at Principe Borghese and Izmir. It appears that Principe Borghese hybrid registered similar values of transpiration rate both in growth and in flowering phenophase. In contrast, Ruxandra hybrid showed an increased rate of transpiration in the flowering phase, which is corelated with an increased rate of photosynthesis in the same phenophase.

The determinations performed in *the fruiting phenophase* indicated an increase of the transpiration rate in all the studied tomato hybrids as result of fruit formation and growth.

The transpiration increase during this development stage may be corelated to an increased water demand provided as result of the roots absorbtion stimulated by an intense transpiration.

It is notable that Izmir and Ruxandra hybrids registered the highest transpiration rate in flowering and fruiting phases, so a positive correlation with photosynthesis rate was observed.

On the contrary, Principe Borghese hybrid registered the lowest rate of transpiration, an almost constant value (2.33-2.45 μmols $H_2O/m^2/s$) during the monitored phenophases. Also the photosynthesis process was the least intense throughout the research in this hybrid.

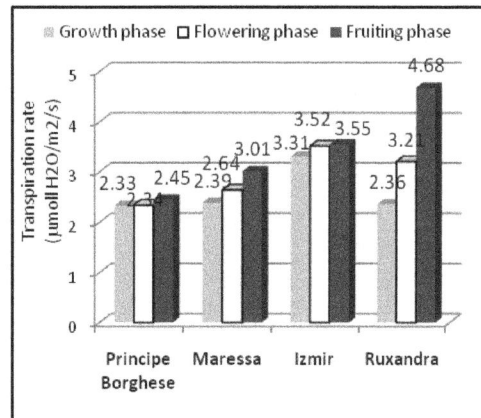

Figure 2. Dynamics of transpiration in the leaves of selected hybrids

The respiration process

The respiration is the only process that provides biochemical energy to achieve the plant growth and development. Tomatoes are climacteric plants, so the respiration process follows a characteristic dynamics: it achieves a maximum in the growth phase, then decreases, but reaches a second maximum (the climacteric maximum) during the maturity phase (Gherghi et al., 2001; Burzo et al., 2005).

Regarding the intensity of respiration process (Figure 3) in the selected tomato hybrids it was noticed high values (2.3-2.22 μmols $CO_2/m^2/s$) at Izmir and Maressa hybrids in *the growth phase*, which are positively correlated to photosynthetic process. The lowest value of respiration rate in this phenophase was registered by the Principe Borghese hybrid (1.49 μmols $CO_2/m^2/s$).

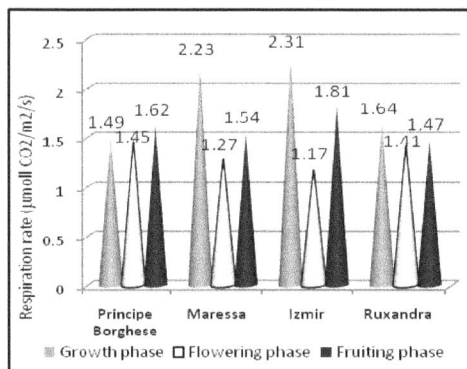

Figure 3. Dynamics of respiration in the leaves of selected hybrids

A decrease of respiration rate was noticed in *the flowering phase* in all the studied tomato hybrids, in according to speciality literature. The reduction rate of respiration was different: by 1.96 times lower in Izmir hybrid, 1.74 times at Maressa, 1.17 times at Ruxandra and insignificant decrease at Principe Borghese hybrid.

In *the fruiting phenophase* a second maximum of respiration process was determined in all the selected hybrids, the highest value (1.81 μmols $CO_2/m^2/s$) being registered at Izmir hybrid, which was characterized by an intense metabolism during the entire research period. A reduced respiration rate was measured at Maressa hybrid (1.54 μmols $CO_2/m^2/s$) and Ruxandra hybrid (1.47 μmols $CO_2/m^2/s$).

CONCLUSIONS

It has been noticed that dynamics of the physiological processes varies depending on the hibrid type and on the development phenophase.

The photosynthesis process follows an ascending evolution during the research period: both the flowering and the fruiting stage stimulate the photosynthesis, which registered increased values compared to the one determined in growth phenophase in all the studied hybrids.

The transpiration process is positively correlated to the photosynthesis, so it increased constantly, but in different rhythm during the development phenophases in all the studied hybrids.

The respiration process follows the dynamics characteristic for climacteric plants: it achieved a maximum in the growth phase, decreased in the flowering phase, but reached a second maximum (the climacteric maximum) in the fruiting phase.

Among studied hybrids, Izmir hybrid was noted with a high biological potential, given the increased values of physiological parameters during the monitored research period, which indicate an intense metabolism.

REFERENCES

Burzo I., Delian E., Dobrescu A., Voican V., Bădulescu L., 2004. Fiziologia plantelor de cultură, Vol. I. Procesele fiziologice din plantele de cultură, Editura Ceres

Burzo I., Dobrescu A., 2005. Fiziologia plantelor, 2005, Ed. Elisavaros, Bucuresti

Gherghi A., Burzo I., Mărgineanu L., Bădulescu L., 2001. Biochimia și Fiziologia Legumelor și Fructelor, Editura Academiei Române

Hetherington S.E., Smillie R.M., Davies W.J., 1998. Photosynthetic activities of vegetative and fruiting tissues of tomato. J. of Exp. Botany, 49(324): 1173–1181

Islam M.T., 2011. Effect of temperature on photosynthesis, yield attributes and yield of tomato genotypes. Int. J. Expt. Agric. 2(1):8-11

Ludwig L.J., Withers A.C., 1984. Photosynthetic response to CO_2 in relation to leaf development in tomato. Adv. in Photosynthetic Res. 4, 217-220

Nour V., Trandafir I., Ionica M.E., 2013. Antioxidant Compounds, Mineral Content and Antioxidant Activity of Several Tomato Cultivars Grown in Southwestern Romania. Not. Bot. Horti. Agrobo., 41(1):136-142

Radzevičius A., Karklelienė R., Viškelis P., Bobinas Č., Bobinaitė R., Sakalauskienė S., 2009. Tomato (*Lycopersicon esculentum Mill.*) fruit quality and physiological parameters at different ripening stages of Lithuanian cultivars. Agr. Res. 7, 712–718

Schwarz D., Kläring H.P., 2002. Growth and Photosynthetic Response of Tomato to Nutrient Solution Concentration at Two Light Levels. J. Amer. Soc. Hort. Sci. 127(6):984–990

Zhu J., Liang Y., Zhu Y., Hao W., Lin X., Wu X., Luo A., 2012. The interactive effects of water and fertilizer on photosynthetic capacity and yield in tomato plants. Austr. J. of Crop Science 6(2):200-20

24

FOLIAR BIOACTIVE TREATMENTS INFLUENCE ON EGGPLANTS SEEDLINGS

Elena DOBRIN[1], Gabriela LUȚĂ[1] , Evelina GHERGHINA[1], Daniela BĂLAN[1], Elena DRĂGHICI [1]

[1]University of Agronomic Sciences and Veterinary Medicine of Bucharest, 59 Marasti Blvd, District 1, Bucharest, Romania

Corresponding author email: glutza@yahoo.com

Abstract

Eggplants (Solanum melongena) is known for its weak root system and for the particular sensitivity towards different stressors such as heat, water stress, nutritional stress. Therefore, is extremely important to obtain eggplants seedlings with balanced growth and development, but especially with a strong root system and increased capacity of adaptation to different conditions of stress. This paper presents the results of some researches related to foliar bioactive substance treatment of eggplants seedling with Spraygard 1%, Razormin 0.1% and BAC Foliar Spray 0.3%. The treatments were performed in two distinct stages of development: at one, respectively three weeks after the seedlings transplantation. Analysis of the acquired data indicated that 0.1% Razormin treatment showed the best results in obtaining of eggplants seedlings with a strong roots systems and a good development. These results are supported by physiological and biochemical processes, which were intensely expressed at the plants in this experimental variant. Also good results in terms of quality seedlings were obtained when treated with BAC Foliar Spray 0.3% and in the case of simple application of adjuvant Spraygard 1%.

Key words*: eggplants, foliar treatments, photosynthesis, root, transpiration.*

INTRODUCTION

In recent years, it was important to reconsider treatments with bioactive substances used in horticultural practice in the context of integrated horticulture development concepts. This comes to align vegetable production in our country to the European Community directives regarding the decrease of environmental pollution through the horticultural technologies, knowing that the horticulture system involves highly intensive cultivation technology which frequently endanger their integrity and the food security.

Vegetables species like eggplant, grown in the temperate climate zone using transplants, are subjected to environmental stress which limits seedlings growth, crop productivity and quality (Sękara et al., 2012). A common consequence of the environmental stress is the increased production of toxic compounds, especially reactive oxygen species produced as result of oxidative metabolism in chloroplasts, mitochondria and peroxisomes (Kim et al., 2004). Previous researches reported that it is absolutely necessary to accomodate the seedlings

with variable stress condition which they will cope in the cultivation place so that allow the achievement of "stress memory" (Jennings and Saltveit, 1994; Knight et al. 1996; Mangrich and Saltveit, 2000; Sękara et al., 2012).

In our country, similar researches have been made on pepper and tomato seedlings that have benefited from fertilization with Razormin and Cropmax, or of treatment with Spraygard (Chilom et al., 2000, Bălan et al., 2014; Dobrin et al., 2014).

The research reported in this study were performed on eggplant (*Solanum melongena*), a main specie of vegetables grown in the field, the summer-autumn crop established by seedlings. It were used different growth regulators (Razormin, Spraygard, BAC Foliar spray, Bio Roots) as foliar treatments on eggplant seedlings and the comparative results were studied.

MATERIALS AND METHODS

The experiments established in 2015, April - June, aimed to test the action of fertilizers Razormin and BAC Foliar Spray and of the universal adjuvant Spraygard on growth and development of eggplants seedlings, in order to be recommended to the seedlings producers as supportive treatment of the growth rate and to improve metabolism seedlings. This could lead to shortening their age, avoidance pests and diseases attacks by strengthening the immune system of plants with implications for reduction of production cost.

The experiment was installed into an experimental greenhouse of the Hortinvest Research Centre – USAMV Bucharest.

Spraygard is a complex product that acts as safener, penetrant, dispersant, creates adhesion of the treatment solutions on the leaves. Spraygard adjuvant has an unique formula in a single coating based on the synthetic resin that is "environmentally friendly" and the polymer di-1-p-menthene and ethoxylated alcohol by applying it on the plant and on its leaves forms a pellicle that persists 2 days up to 2 weeks, having as a side effect the reduction of perspiration and, therefore, a better water management within the plant. This fact causes the physiological chain reactions whose results are being expressed by increasing the plant resistance to stress factors such as the drought and the cold. The effect of reducing transpiration recommends applying the product strictly on the leaves.

Razormin is an environmentally friendly bio-stimulating product, which determines a rooting effect. Their chemical composition is complex and balanced, so that induces mainly a root system development, than the development of vegetative part through cell division. It contains free amino acids and polysaccharides, which stimulate the nutrients absorption, leading to the further development of plant.

BAC Foliar is a foliar organic nutrient which stimulates chlorophyll production in the leaves.

We established a monofactorial experiment with 4 variants, considering application of bioactives substances Sparygard 1%, Razormin 0.1% and BAC Foliar Spray 0.3% on egg-plants seedlings (Pana Corbului – an appreciate Romanian varieties) in two distinct stages: at one, respectively three weeks after the seedlings transplantation (7 May and 22 May). The experimental variants were: V1–untreated seedlings; V2–seedlings treated with 1% Spraygard; V3–seedlings trated with 0.1%

Razormin; V4–seedlings treated with BAC Foliar Spray 0.3%. The experiment was installed using the block method in linear alignment with 4 repetitions. The total number of plants in the experiment was 240, each variant containing 60 plants, with 15 plants per repetition.

Sowing was made directly into alveolar pallets (alveolar ø = 5 cm) on April. Because heat and water were optimal provided, mass emergence of seedlings occurred after 12 days. The eggplants seedlings transplanting was done on April 26, in large plastic pots (400 ml) filled with professionally nutrient substrate KEKKILA BP 75% + 25% perlite. It has used this type of pots, with high volume, to counteract any imbalances that may arise in terms of installing a high or even excessive thermal and hydric conditions in the production are, specific at this time of year. During the growth period specific agrotechnics for seedling production was applied: daily ventilation, watering, weeding weeds. The temperature was kept at 22–24 °C to 28 °C at day and 18–20 °C at night. A treatment with CE Bravo 0.2% was made in order to prevent seedlings fall and also to avoid a *downy mildew* attack.

Observations and measurements of plant growth were made during the development of experiments in different stages: a week and respectively five weeks after transplantation.

Observations and measurements were made on seedlings growth, as follow:

• **biometric parameters of seedlings**: plant height; the number of true leaves; weight of aerial vegetative unit; seedlings total weight; root weight and volume;

• **measurements of the main physio-logical processes intensity** (photosynthesis, transpiration, stomatal conductance) at the end of the experiment. We used the LC pro+ photosynthesis system. The measurements were performed on the active leaves located in the middle third part of the plant;

• **determinations of the assimilatory pigments content** in the active leaves: *chlorophyll* and *carotenoid pigments* were extracted in 80% acetone and determined spectrophotometrically (wavelenghts 663 nm, 647 nm and 480 nm) using the extinction coefficients and equations described by

Schopfer (1989). The results were expressed in mg/100 g fresh weight.

RESULTS AND DISCUSSIONS

The results of the analysis of the first stage (10 days after the first treatment) are shown in Table 1. Eggplant seedlings have reacted differently to the treatments applied. Plant height is quite different, from 12.4 cm (V1) to 13.6 cm from V4 - fertilized with BAC Foliar Spray, 0.3%. Also in this variant it was recorded the highest number of leaves 5, while at the untreated variant was of 4.2 leaves. The frequency is 0.34 leaves/cm PA to the V1 and 0.38 leaf/cm PA for V2. and V4.

Table 1. Growth and development of eggplant seedlings 10 days after the first fertilization

Variant	Plants height HPA (cm)	No. of leaves	Leaves frequency (nr/ cm HPA)
V₁	12.4	4.2	0.34
V₂	12.8	4.8	0.38
V₃	13.4	4.8	0.36
V₄	13.6	5.0	0.37

Context analysis at this time points out that the application of different treatments have a defining influence on plant growth and development (R^2 ≥ 0.9692 for plant height and R^2 ≥ 0.8 for the number of leaves formed) and allows placement variant treatment BAC Foliar Spray, 0.3% on top position, closely followed by treatment with Razormin 0.1% (figure 1).

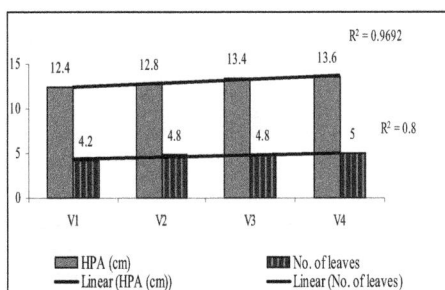

Figure 1. Influence of applied treatment on growth of eggplant seedlings ten days after treatments

In order to determine the overall effect of the treatments program applied on the eggplant seedlings were made observations and

measurements also two weeks after application of the second treatment. The results obtained are shown in Tables 2 and 3, respective in Figures 2, 3 and 4.

Table 2. Growth of eggplant seedlings at two weeks after the second treatment

Variant	No. of leaves	Plants height HPA (cm)	Roots length HR (cm)	Plants total lenght HT(cm)	Leaves frequency (nr./cm HPA)
V₁	5.4	21.8	10.6	32.4	0.25
V₂	6.4	24.6	12.2	36.8	0.26
V₃	7.4	24.8	17.2	42.0	0.30
V₄	6.6	23.2	16.0	39.2	0.28

Applied treatment program determined differences regarding on the growth of eggplant seedlings. Analysis of the results on the growth of seedlings showed that the best option working was V3 - Razormin 0.1%. In this variant plants have achieved the best and balanced growth, all indicators analyzed had the higher values (7.4 leaves formed, 24.8 cm plant height, 17.2 cm root length, total length 42 cm plant; 0.3 frequency leaves). In contrast, V1 untreated produced the smallest increase, all the analyzed indicators registering the lowest values (5.4 leaves formed, 21.8 cm plant height, 10.6 cm root length, total length of 32.4 cm plants, leaves frequency 0.25).

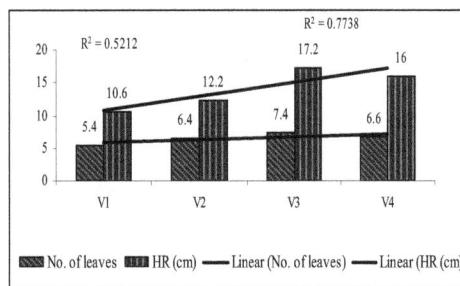

Figure 2. Influence of applied treatment on eggplant seedlings growth two weeks after the second treatment

As can be seen from Figure 2 and 3, schedule treatments with bioactive substances exert a greater influence on the growth of eggplant seedlings, respectively, a very significant influence on the growth of roots (R^2 = 0.7738) and a significant one on the number of leaves and on frequency leaves (R^2≥0.52).

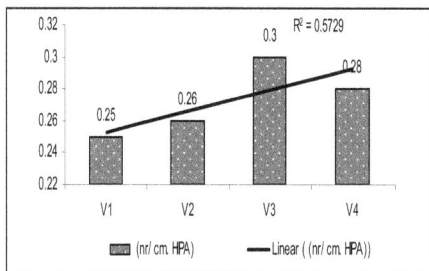

Figure 3. Influence of applied treatment on the leaves
frequency two weeks after the second treatment

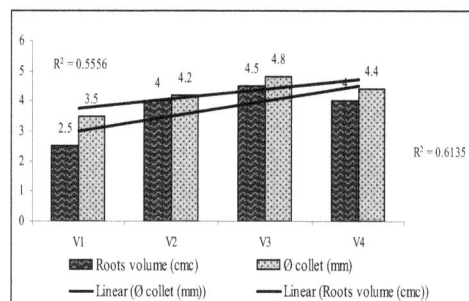

Figure 4. Influence of applied treatment on the roots
weight two weeks after the second treatment

Taken together the results obtained for eggplant
seedlings morphometry, we estimate that the
most balanced variant is V3 (fertilized with
0.1% Razormin) regarding on plants growth.
Developing of eggplant seedlings two weeks
after the second treatments was quantified by
various indicators of weight and volume and by
diameter of collet. The obtained results (Table
3; Figure 4 and 5) regarding eggplant mass
ratio highlights two situations:

    1. at V3 variant all indicators recorded the
highest values compare to the other variants
(mass root 4.5 g; 10 g total mass, volume root
4.5 cm^3; 4.8 mm collet diameter), excluding the
aerial part mass (5.5 g);

    2. in contrast, V1 variant recorded the
lowest values (mass root 2 g; 8 g total mass;
volume root 2.5 cm^3; 3.5 mm collet diameter),
excluding the aerial part mass (6 g).

Good results have also recorded the seedlings
treated with BAC Foliar spray and Spraygard.
A strong influence of the treatment on the
development of the root system and of the
collet diameter was noticed.

Table 3. Developing of eggplant seedlings at two weeks
after the second treatments

Variant	Aerial part mass (g)	Roots weight (g)	Total weight (g)	Roots volume (cm^3)	Ø collet (mm)
V$_1$	6.0	2.0	8.0	2.5	3.5
V$_2$	5.5	4.0	9.5	4.0	4.2
V$_3$	5.5	4.5	10.0	4.5	4.8
V$_4$	5.5	4.0	9.5	4.0	4.4

The results of the physiological measurements
performed on the experimental variants are
shown in Table 4. As can be seen, the leaf
temperature was relatively constant (27.2–27.6
°C) and light intensity (Q) registered the value
of 1280-1360 mmol/m^2/s.

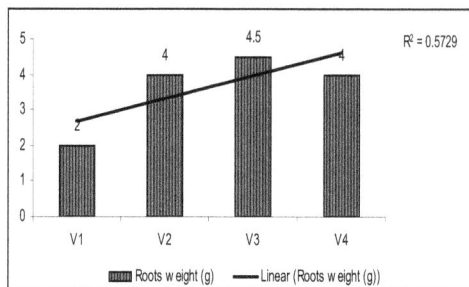

Figure 5. Influence of applied treatment on the roots
volume and the collet diameter two weeks after the
second treatment

Table 4. Physiology of the eggplant seedlings two weeks
after the second treatment

Var.	A [μmol/ m^2/s]	E [μmol/ m^2/s]	A/E	GS [μmol/ m^2/s]	Leaf temp.T [^{0}C]	Q [μmoli/ m^2/s]
V$_1$	9.95	1.84	5.41	0.07	27.2	1280
V$_2$	6.35	0.84	7.56	0.045	27.6	1280
V$_3$	8.34	1.80	4.63	0.06	27.6	1360
V$_4$	9.08	1.57	5.78	0.06	27.2	1360

The results analysis revealed that V1 untreated
variant recorded the highest values
(Photosynthesis rate A = 9.95μmol/m^2/s;
Transpiration rate E = 1.84 μmol/m^2/s;
Stomatal conductance Gs = 0.07; A/E = 5.41)
for all studied parameters. This intense
physiological activity is not supported by plant
growth and can be explained only by the theory
that nutrition, regime of water and the
temperature may be major factors of stress for
seedlings of eggplant (Acatrinei, 2010; Sekara
et al., 2012). The intensify of physiological
processes without correlation with translocation
and accumulation of photoassimilated
substances is, in fact, the response of plants to
the action of stressors whose intensity action
does not endanger the life of plants. On the
overall results can be noted V4 (BAC Foliar

spray) as the most balanced variant regarding physiological activities. It was also noticed V2 - Spraygard, which amid of low rates of transpiration and stomatal conductance, recorded very good photosynthetic yields.

Table 5. Assimilatory pigments of the seedlings leaves

Var.	Assimilatory pigments (mg/100 g)		
	Chlorophyll a	Chlorophyll b	Carotenoids
V_1	102.43	50.42	5.70
V_2	112.00	38.98	5.27
V_3	136.50	47.97	5.41
V_4	142.67	36.35	4.35

Biochemistry leaves are somewhat contradictory to physiology in the sense that an increased rate of photosynthesis is not necessarily correlated with an increased content of chlorophyll pigments. Thus, V1 untreated, recorded the lowest content of chlorophyll pigments (chlorophyll 102.43 mg/100 g, 50.42 mg/100 g chlorophyll b) but, due to an increase in photoprotection status, had the highest content in carotenoids pigments (5.7 mg/100 g). For other variants studied, the results were correlated with physiology and morphometry. It was again emphasized the V3, with the highest content of chlorophyll pigments and a very good photoprotective activity. V4 occupied second place, which was given mainly to synthesis of chlorophyll a (142.67 mg/100 g).

CONCLUSIONS

In horticultural tehnology were applied treatments with bioactive substances as a frequence practice for accelerating or inhibiting the growth of vegetable seedlings or plant or as a life support under different stressful conditions. In the first stage of analysis the context results shows that treated variants have superior biometrics comparing to the control untreated variant. BAC Foliar Spray 0.3% highlights as variant in which seedlings have realised the most favorable growth.
At the second moment of analysis V3 - fertilized seedlings Razormin 0.1% it was noticed as the best option working variant. In this variant plants achieved the best and balanced growth, all indicators analyzed having higher values (7.4 leaves formed, 24.8 cm plant height, 17.2 cm root length, total length 42 cm plant; 0.3 frequency leaves). Treatments

program applied of eggplant seedlings exerted at least a significant influence on the growth of eggplant seedlings, of roots, of the number of leaves and of frequency leaves.
Mass ratio analysis reveals that at V3 variant all indicators recorded the highest values compare to the other variants excluding the aerial part mass
Biochemical and physiological analysis also revealed V3 with the highest content of chlorophyll pigments and a very good photoprotective activity. V4 occupied second place, which was given mainly to synthesis of chlorophyll a.
In conclusion, the analysis of the acquired data allow to remark that treatment with Razormin 0.1% led to obtaining of eggplant seedlings with high quality and strong roots. Very good results in terms of quality seedlings were obtained when treatment with BAC Foliar Spray 0.3% was used and also with simple application of adjuvant Spraygard 1%.

REFERENCES

Bălan D., Dobrin E., Luță G., Gherghina E., 2014. Foliar fertilization influence on pepper seedlings. Analele Universității din Craiova, Vol. XIX (LV) – 2014, pag. 27-32.

Chilom P., Bălașa M., Dinu M., Poștaliu G., Spirescu C., 2000. Cropmax, îngrășământ biologic complex cu stimulatori de creștere pentru fertilizarea foliară. Rev. „Sănătatea plantelor", nr. 6, Bucuresti.

Dobrin E., Roșu M., Drăghici E., Gherghina E., 2014. Research on the influence of treatment with Spraygard of quality seedlings. Analele Universității din Craiova, Vol. XIX (LV), pag. 161-166.

Jennings P., Saltveit M. E., 1994, Temperature and chemical shocks induce chilling tolerance in germinating Cucumis sativus (cv. Poinsett 76) seeds. Physiol. Plant. 91:703-707.

Kim K., Portis A.R., 2004. Oxygen-dependent H_2O_2 production by Rubisco. FEBS Letters, 571, 124-128.

Knight H., Trewavas A.J., Knight M.R., 1996. Cold calcium signaling in Arabidopsis involves two cellular pools and a change in calcium signature after cold acclimation. Plant Cell, 8:489-503.

Mangrich M.E., Saltveit M.E. , 2000. Effect of chilling, heat shock and vigor on the growth of cucumber (Cucumis sativus) radicles. Physiol. Plant 109:137-142.

Schopfer P., 1989. Experimentelle Pflanzenphysiologie. Berlin, Springer-Verlag, p. 33-35.

Sękara A., Bączek-Kwinta A., Kalisz A., Cebula S., 2012. Tolerance of eggplant (Solanum melongena L.) seedlings to stress factors. Acta Agrobotanica Vol. 65 (2): 83–92.

25

EVALUATION OF PARENTAL FORMS AND HYBRID POPULATIONS DESCENDING FROM TOMATOES, FOLLOWING HEAT RESISTANCE AND PRODUCTIVITY

Nadejda MIHNEA[1]

[1]Institute of Genetics, Physiology and Plant Protection, Academy of Sciences of Moldova, 20 Pădurii street, MD-2002, Chişinău, Republic of Moldova, e-mail: mihneanadea@yahoo.com

Corresponding author email: mihneanadea@yahoo.com

Abstract

Climate change has a negative impact on agriculture. In the last decade, drought and high temperatures have become more frequent with strong negative effects on crop productivity. Selection based on resistance to extreme temperatures now becomes the actual objective because in some years, the air temperature reaches 35-45°C during the flowering stage of the day, which considerably reduces fruits setting and the yield per hectare. The aim of our research was to test the level of heat resistance of genotypes and hybrid populations descending from tomatoes to select the forms with high resistance level. As a result of research we found that the highest resistance to heat was manifested by genotypes of F_3 generation Mihaela x Irisca (93.9%) and F_3 Maestro x Irisca (82.9%) and the lowest - by the variety Maestro (31. 2%). The genotypes with increased resistance can be used subsequently in research for genetic breeding for resistance to high temperatures.Testing of the selected material based on the characters complex, including heat resistance, demonstrated the possibility of creating new forms of tomato combining productivity with high air temperatures resistance.

Key words: tomatoes, intra-specific hybridization, breeding, resistance, heat

INTRODUCTION

Tomatoes are one of the most common vegetable crops in Moldova and worldwide because of high nutritional value of its fruits, both, for fresh consumption and many types of processed products.

In Moldova, permanently, climate change has had a negative impact on agriculture. We can mention that in the past decade drought and high temperatures have become more frequent, with strong negative effects on crop productivity. Selection based on resistance to extreme temperatures now becomes the actual objective because in some years the air temperature reaches 35-45°C during the day at the blossoming stage, which considerably reduces fruits setting and the yield per hectare. Therefore, creating tomato genotypes with resistance to extreme environmental factors and high productivity and quality indicators presents a serious improvement (Mihnea et al., 2010; Mihnea, 2011; Mihnea et al., 2005; Moldovanu et al., 2000. Sato et al., 2000; Saltanovici et al., 2003; Saltanovici et al., 2012). According to the author V.L Erşova

(1979) the air temperature is of 30-33°C for most tomato varieties, the pollen losses its fertility, fecundity is compromised, the flowers fall, growth either stops or interrupts and the intensity of photosynthesis decreases. The harmful impact of high temperatures is intensified under deficit of soil water. It has been established that the optimum temperature of pollen germination is from 22 to 26°C and fruit development - from 20 to 24°C.

Positive effects of breeding can be achieved by using a sufficient number of genotypes specially selected for certain agro-environmental areas and, at the same time, taking into account the considerable variability of the plants cultivation conditions. Contemporary breeding demonstrates the need to create lines, varieties and hybrids with high environmental resistance. The importance of the adaptive breeding to create varieties combining resistance and stressogenic factors with high productivity factors has been recognized by many researchers (Pivovarov et al, 1990; Kilchevsky, 1997; Zhuchenko, 2005).

The aim of our research was to test the level of heat resistance of genotypes and hybrid populations for selecting forms descending from tomatoes with the high level of heat resistance.

MATERIALS AND METHODS

The experiments were conducted in the year 2011 under field conditions in the experimental plot of the IGFPP. As the initial material for the planned research were used genotypes selected from F_3 hybrid generations using a character complex of 4 combinations and backcrossing hybrids obtained on the base of intra-specific hybridization of Maesrto x Irisca, Maestro x Dwarf Moneymaker, Maestro x Dwarf Moneymaker, Mihaela x Iriska. Field experiments were conducted in triplicate in randomized blocks of seedlings cultivation without irrigation. The sowing took place in greenhouses in the first decade of April according to the scheme 7 x 10 cm and field planting - in the scheme of 70 x 30 cm. Field planting was performed in the second decade of May, and harvesting was done gradually (4-6).
Caring for and tomato growing were performed in accordance with agro- technical norms adopted in Moldova. High temperature resistance of genotypes was evaluated according to methodological recommendations (Ivakin, 1979) based on plant growth capacity after their exposure to high temperatures (43°C) for 6 hours. The data obtained were statistically processed using the software STATISTICS 7.
Graphical representation, tabular and textual, was performed through the Microsoft Office and Microsoft Excel software.

RESULTS AND DISCUSSIONS

By testing the response of 4 parental forms and 43 phenotypes of tomatoes selected from four hybrid combinations and backcrossing hybrids obtained on the base of intra-specific hybridization of Maesrto x Irisca, Maestro x Dwarf Moneymaker, Maestro x Dwarf Moneymaker, Mihaela x Irisca hybrids under the impact of high temperatures (43° C) it was established that genotypes/populations selected for two years under field conditions manifested different responses to the heat. The data obtained on the response of tomato hybrid populations at elevated temperatures

demonstrated that in some hybrid combinations F_3, F_2BC there were recorded values of high resistance of hybrids obtained in comparison with the genitors, while in some value combinations there were less average parents. The result of evaluating genotypes / populations on heat resistance is shown in Figure 1. Analysis of these forms under heat resistance showed that variability from 31.2 to 93.9% was within the limits. As the data show, a high resistance was demonstrated by genotypes 4, 6, 7, 8 selected from F_2 hybrid combination Maesrto x Irisca, genotypes 11, 13 from the backcross combination F_1BC (Maesrto x Irisca) x Maestro, genotype 15 from combination F_1BC (Maestro x Irisca) x Irisca. According to the degree of resistance to temperatures indicated by genotypes selected from combinations in which as the paternal form was used the variety Dwarf Moneymaker, there were found two genotypes of 21 (5 - F_3 Maestro x Dwarf Moneymaker and 11 -F_1 BC (Mihaela x Dwarf Moneymaker) x Dwarf Moneymaker]) that demonstrated resistance of 77.9% and 60.3%. In terms of the heat response of genotypes selected from the combination Mihaela x Irisca, it was seen that that genotypes 5, 6, 9 (Fig. 1C) demonstrated the resistance of 63.1; 93.9 and 61.4%, respectively. It should be mentioned that most simple and backcrossed hybrid populations showed high values of the resistance level of plants. However, hybrid populations created with participation of Irisca variety had the highest indices of the examined character.
Concerning another examined quantitative indicators - the seedling length that can be considered both, as the genotype peculiarity and the index of resistance to high temperature and it is evident that it differs much from genotypes included in the study (Fig. 2).
Plants length under optimum conditions was within the limits of 70.0 ... 114.2 mm while under stressful conditions - 24.2 ... 66.6 mm in case of the combination Maesrto x Irisca whose parents showed pronounced differences of the analyzed index (24.2 and 41.0 mm); it was found that most offspring had much higher values than the best parents, with the exception of genotypes 10, 16 in which the plant length was 33.9 and 33.2 mm. The offspring from the hybrid combination Mihaela x Irişca (Figure 2

C) showed a higher level of plats growth and was within the limits from 46.6 to 66.6 mm. In genotypes selected from the combination Maestro x Dwarf Money Maker (Figure 1) the

plants length under stress was greater than in the parental forms while in combination of Mihaela x Dwarf Moneymaker, only one descendent ceded the best parent.

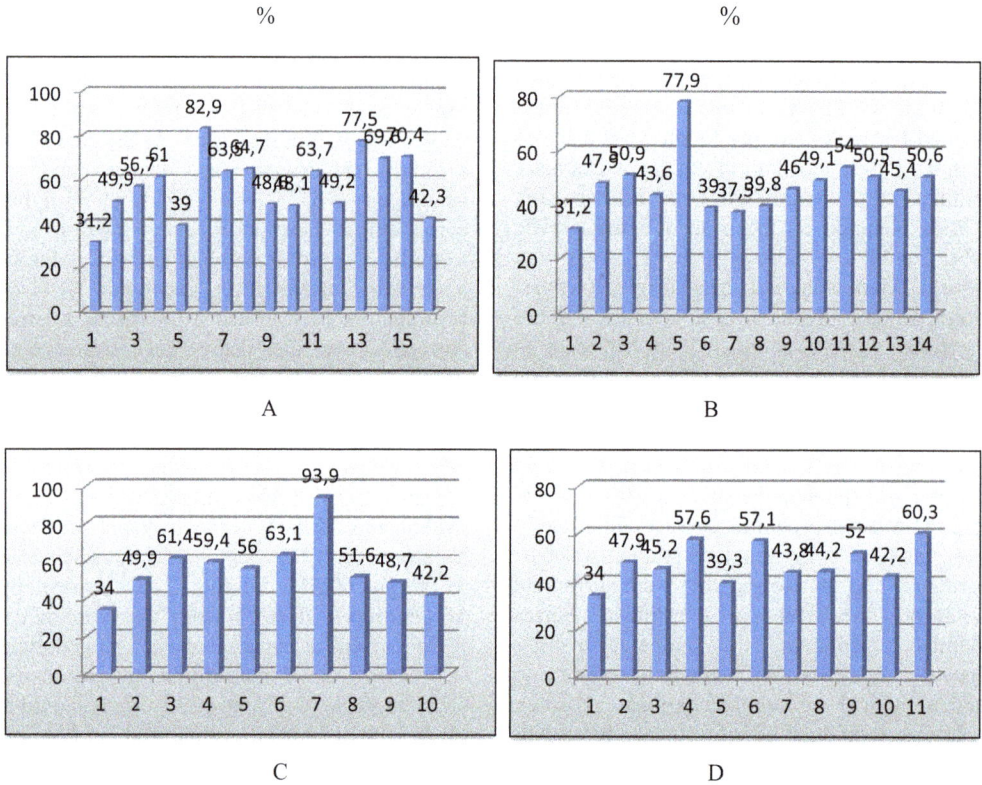

%

%

A

B

C

D

Figure 1. Evaluation of tomato genotypes selected from intra-specific combinations on heat resistance in comparison with parental forms
A. 1 – Maestro, 2 – Irişca, 3-9 – F_3 (Maesrto x Irişca), 10-13 – F_2 BC (Maesrto x Irişca) x Maestro, 14-16 – F_2 BC (Maesrto x Irişca) x Irişca.
B. 1 – Maesrto, 2 - Dwarf Moneymaker, 3-10 – F_3 (Maestro x Dwarf Moneymaker), 11-12 – F_2 BC (Maestro x Dwarf Moneymaker) x Maestro, 13-14 – F_2 BC (Maestro x Dwarf Moneymaker) x Dwarf Moneymaker.
C. 1 - Mihaela, 2 – Irişca, 3-7 – F_3 (Mihaela x Irişca), 8-9 - F_2 BC (Mihaela x Irişca) x Mihaela, 10 – F_2 BC Mihaela x Irişca) x Irişca.
D. 1 – Mihaela, 2 - Dwarf Moneymaker, 3-8 – F_3 (Mihaela x Dwarf Moneymaker), 9-10 – F_2 BC Mihaela x Dwarf Moneymaker) x Mihaela, 11 – F_2 BC (Mihaela x Dwarf Moneymaker) x Dwarf Moneymaker.

Under the influence of temperature of 43^0 C the repression of plant growth occurred in all analyzed forms, whose values were presented in genotypes of the combination MaestRo x Irisca - 31.8; -46.4; -42.5; - 27%; -38.9; -44.0; -41.2; -49.2; -34.9; - 43.9; -29.5; - 53.3 -54.6; -56.0%, while in the combination Mihaela x Irisca -44.; -25 4; -37.9; - 22.8; -31.5; -24.5; -43.9; -57.1% compared with the check. In the combination of Maestro x Dwarf Moneymaker

the repression of plant growth was within the limits of 36.9 ... 65.5, while in Mihaela x Dwarf Moneymaker - 42.3 ... 57.9%. Testing of the selected material on the base of the parent sporophyte heat resistance revealed the forms possessing a lower depression of the plants under the thermal stress and high resistance to heat. They were included in the process of further improvement.

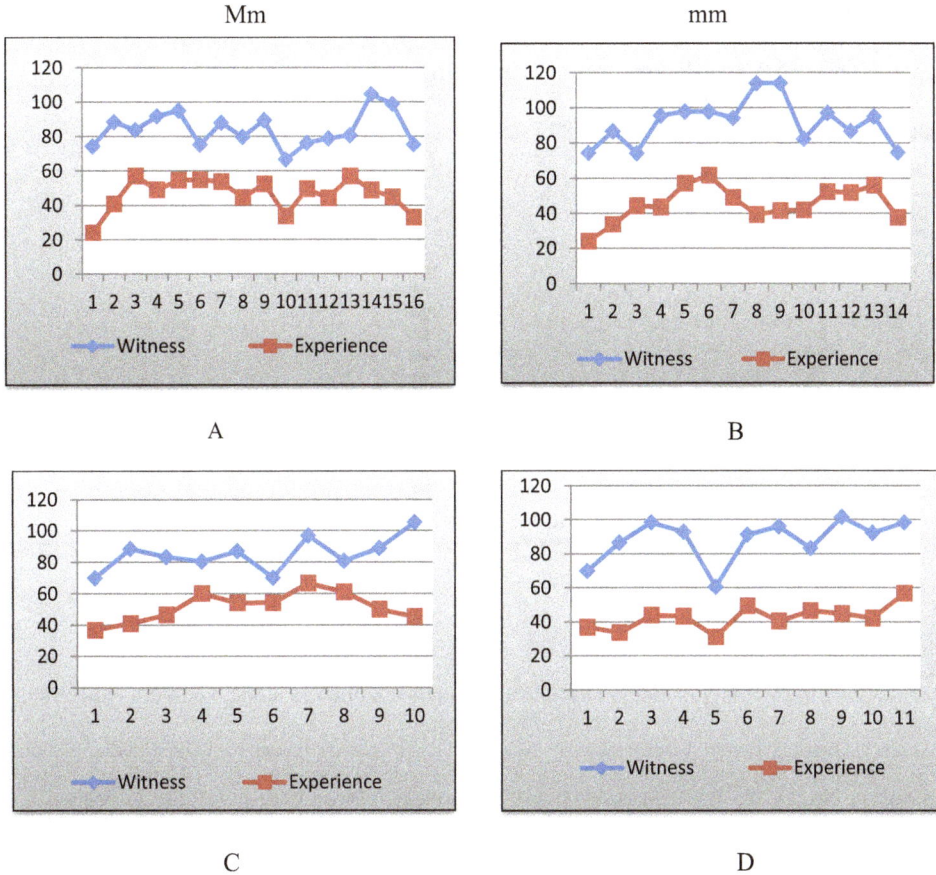

Figure 2. Tomato plant length under optimal conditions (25^0 C) and stress ($43\ ^0$ C)

A. 1 – Maestro, 2 – Irişca, 3-9 – F_3 (Maesrto x Irişca), 10-13 – F_2 BC (Maesrto x Irişca) x Maestro, 14-16 – F_2 BC (Maesrto x Irişca) x Irişca.

B. 1 – Maesrto, 2 - Dwarf Moneymaker, 3-10 – F_3 (Maestro x Dwarf Moneymaker), 11-12 – F_2 BC (Maestro x Dwarf Moneymaker) x Maestro, 13-14 – F_2 BC (Maestro x Dwarf Moneymaker) x Dwarf Moneymaker.

C. 1 - Mihaela, 2 – Irişca, 3-7 – F_3 (Mihaela x Irişca), 8-9 - F_2 BC (Mihaela x Irişca) x Mihaela, 10 – F_2 BC Mihaela x Irişca) x Irişca.

D. 1 – Mihaela, 2 - Dwarf Moneymaker, 3-8 – F_3 (Mihaela x Dwarf Moneymaker), 9-10 – F_2 BC Mihaela x Dwarf Moneymaker) x Mihaela, 11 – F_2 BC (Mihaela x Dwarf Moneymaker) x Dwarf Moneymaker.

Evaluation promising forms under field conditions in respect of productivity revealed a rather high variability of both the total harvest (Figure 3 A) and the share of market fruit (Figure 3 B).The total harvest of the initial forms ranged from 44.9 t / ha (variety Irişca) to 55.2 (variety Dwarf Moneymaker), the rate of market fruit was 87.0 ... 97.0%. The total harvest in hybrid combination F_3 (Maestro x Irishka) and F2 BC (Maestro x Irişca) x Maestro was 53.0 and 55.7 t / ha, much higher than of the parent with high values, variety Maestro (48.2 t / ha). The hybrid combinations F_2 BC (Maestro x Irişca) x Irişca the harvest was at the level of the best parent, and the hybrid combinations where for crosses were used varieties Maestro and Dwarf Moneymaker, lines showed a lower harvest with the exception of the combination F_2BC (Maestro x Dwarf Moneymaker) x Maestro. In combinations derived from crosses of varieties Mihaela and Irisca, Mihaela and Dwarf Moneymaker the total harvest was lower than the best parent except combination F3 (Maeda x Dwarf Moneymaker) for which the total harvest was much higher than of the parent with high values. After evaluating the market rate of fruit it was found that hybrid combinations showed a variability of 87.8 ...

97.0%. A low market rate of fruit was recorded for the hybrid combinations F₂BC (Mihaela x Irişca) x Mihaela (89,8%).

t/ha

%

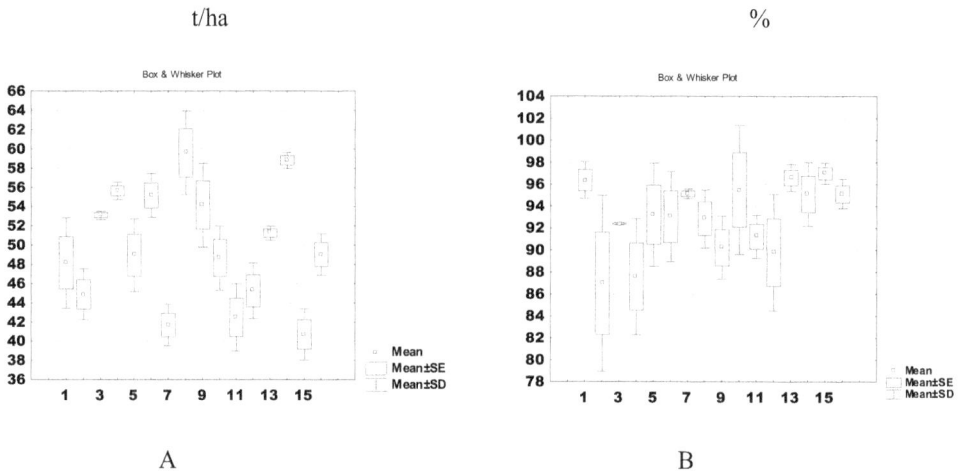

A

B

Figure 3. Productivity of tomato varieties and lines obtained as a result of intra –specific hybridization, year 2011

1 – Maestro, 2 – Irişca, 3 - Dwarf Moneymaker, 4 – Mihaela, 5 – F_3 (Maestro x Irişca); 6 – F_2 BC (Maestro x Irişca) x Maestro, 7 – F_2 BC (Maestro x Irişca) x Irişca, 8 – F_3 (Maestro x Dwarf Moneymaker), 9 – F_2 BC (Maestro x Dwarf Moneymaker) x Maestro, 10 – F_2 BC (Maestro x Dwarf Moneymaker) x Dwarf Moneymaker, 11 – F_3 (Mihaela x Irişca), 12 – F_2BC (Mihaela x Irişca) x Михаела, 13 – F_2 BC (Mihaela x Irişca) x Irişca, 14 – F_3 (Mihaela x Dwarf Moneymaker), 15 – F_2 BC (Mihaela x Dwarf Moneymaker) x Mihaela, 16 – F_2 BC (Mihaela x Dwarf Moneymaker) x Dwarf Moneymaker

CONCLUSIONS

The highest heat resistance was found in the genotypes of the F_3 generation (Mihaela x Irisca (93.9%) and F_3 Maestro x Irisca (82.9%), and the lowest in the variety Maestro (31.2%).

The genotypes tested showed differential response to heat and can be used later in the research of genetic resistance improvement to high temperatures.

Testing of the selected material based on a character complex, including heat resistance, demonstrated the possibility of creating new tomato forms that combines productivity with resistance to high air temperatures.

REFERENCES

Erşova V., 1978. Cultivarea tomatelor în câmp deschis neprotejat. Ştiinţa, Chişinău.
Ivakin A., 1979. Determinarea rezistenţei la arşiţă a culturilor legumicole în baza reacţiei de creştere a plantelor după menţinerea la temperaturi înalte. Indicaţii metodice. Institutul de fitotehnie, Leningrad.
Jucenco A., 2005. Natura genetică a capacităţii adaptive la plantele de cultură. În: Identificarea genofondului plantelor şi ameliorarea, Sank-Petersburg.
Kilicevschi A., Hotîleva L., 1997. Ameliorarea ecologică la plante. Tehnologia, Minsk, 36-101.
Mihnea N., 2005. Crearea materialului iniţial pentru ameliorarea soiurilor rezistente la arşiţă. Starea actuală şi dezvoltarea contemporană a selecţiei şi semenologiei plantelor legumicole. Vol.1, Moscova, 262-264.
Mihnea N., 2010. Potenţialul genetic de rezistenţă la arşiţă al soiurilor şi liniilor valoroase de tomate. Buletinul Academiei de Ştiinţe a Moldovei. Ştiinţele Vieţii; 2(308), Chişinău. 64-69.
Mihnea N., 2010. Valoarea genetico-ameliorativă a formelor de tomate cu port erect. Buletinul Academiei de Ştiinţe a Moldovei. Ştiinţele Vieţii, ediţie specială (Rapoarte prezentate la Congresul al IX-lea al Geneticienilor şi Amelioratorilor), 2(311), Chişinău. 125-131.
Mihnea N., 2011. Evaluarea comparativă a liniilor noi de tomate rezistente la arşiţă. Plante noi netradiţionale şi utilizarea lor în perspectivă. Materialele

Simpozionului IX-a International. Vol. 1, Moscova, 188-190.

Mihnea N., 2011. Linii perspective de tomate cu potențial de productivitate și adaptabilitate la condițiile din Republica Moldova. Buletinul Academiei de Științe a Moldovei. Științele Vieții, nr.2(314), Chișinău: 124-130.

Moldovanu L., 2000. Ameliorarea rezistenței tomatelor la temperaturi extremale. În: Ameliorarea și semenologia culturilor legumicole în secolul XXI. Conferința științifico-practică Internațională, Moscova, 92-93

Pivovarov V., 1990. Probleme de ameliorare ecologică a culturilor legumicole. Horticultura intensivă, Gorchi. 57-63.

Saltanovici T., 2003. Recoltarea gameților, rezistenți la factorii ecologici ai mediului. Buletinul Academiei de Științe a Moldovei. Științele biologice, chimice și agronomice, nr 2. Chișinău: 98-100.

Saltanovici T., 2012. Utilizarea tehnologiilor gametice la testarea genotipurilor de tomate. Intelectus: 25-30.

Sato S., 2000. Physiological factors limit fruit set of tomato (*Lycopersicon esculentum* Mill.) under chronic, mild heat stress. In: Plant, Cell and Environment, vol. 23:719–726.

26

BACKGROUND OF VEGETABLE MULTI- CRITERIA ANALYSIS IN MUNICIPALITY OF KALOYANOVO, BULGARIA

Vera STEFANOVA[1]

[1]Agricultural University of Plovdiv, 12 Mendeleev Blvd, Plovdiv, Bulgaria Corresponding author email: vera.v.stefanova@abv.bg

Abstract

Horticulture is one of the importance parts of Bulgarian agricultural sectors. Crop growing specification requires deep and good knowledge about agricultural development and practices. Advanced research and carefully gathered and analysed information is the first step of the successful result. This publication presents the structure and more necessary data for complex analyzing of environmental factors and their influence on agriculture structure in Municipality of Kaloyanovo, Plovdiv region, Bulgaria. The study area is 347 sq.km and average elevation - 198 m. An overview would show the present state of physiography, climate, crop cultivation, landscape, elevation, labor force, marketing, farmers practices, areas under cultivation, land use, soil type, water resources, road systems, erosion level, the most spread vegetable marketing, demography and socio-economy opportunities. Methodology is based on thematic maps and analysis, creating by GIS application. It aims to identify current status of the most appropriate plants development, land management and changes in land use form. Spatial data and thematic maps present increase and decrease of agricultural availability in the study area. Results obtained will be relate to multi-criteria land suitability evaluation, accordingly to vegetable crop characteristics. This model of land evaluation helps farmers and vegetable growers to realise potentiality of their land parcel and required management procedures.

Key words: multi-criteria analysis, GIS, environmental factors, vegetable crops.

INTRODUCTION

Agriculture is permanent developing structure, connected to the sensible land use and land management. From the beginning of the civilization man has used the land resources to satisfy his needs. The land resources regeneration is very slow while the population growth is very fast, leading to a nunbalance. On a global scale, agriculture has the proven potential to increase food suppliesfaster than the growth of the population (Davidson, 1992). Sustainable vegetable farming system is associated with good practices related to people cantered development, sustainable live lihood, agro-ecological practices, sustainable forestry system, community based natural resources management, participatory policy development, indigenous farming system, fair labour condition, good agricultural practises, equitable access to water and others (Baniya, 2008). Sustainability is the ability of an agricultural system to meet evolving human needs without destroying and, if possible, by improving the natural resource base on which it depends (USAID, 1988). In order to determine the most desirable direction for future development, the suitability for various land uses should be carefully studied with the aim of directing growth to the most appropriate sites.

Necessary of sustainable agricultural development is mentioned by different authors (Stoeva, 2013; Christova et al, 2013; Toskov, 2013; Nikolova, 2013). According to them, it is important to create developing strategies about vegetable production and management. Natural resources have to be ruled and used carefully for good sustainable future yields. Careful planning of the use of land resources is based on land evaluation, which is the process of assessing the suitability of land for alternative land uses (Fresco et al, 1994). Information on land resources is a key to their careful and effective evaluation.

The main purpose of this article is to present the important and necessary environmental and human developing factors for formulating sustainable vegetable developing. Establishing appropriate suitability factors is the construction of suitability analysis. Agricultural evaluation is concerned with the assessment of various land actions and used for specified purposes. It involves an analysis of some basic surveys of geographic situation, soils, land using, demography state, saved areas by lows, road systems, water resources, etc. All data is response with plant requirements. This is mean to present the importance of preliminary environmental assessment of some area and its positive and negative sides for sustainable vegetable developing.

Study area

In Bulgaria it is not so spread and usual, farmers to make a preliminary study of the future cultivated area. This study presents the essence of necessary factors for making good vegetable production. The study area is Municipality of Kaloyanovo. Situated in the north part of the Upper Thracian plain, covering area of 347 sq. km., the municipality is part of the region of Plovdiv and consists of the municipal center Kaloyanovo, and 14 settlements as well - Begovo, Glavatar, Gorna Mahala, Dolna Mahala, Duvanlii, Dalgo Pole, Zhitnitsa, Ivan Vazovo, Otets Paisievo, Pesnopoy, Razhevo, Razhevo Konare, Suhozem and Chernozemen. Up to the present moment the population of the municipality amounts to 12800 people. Municipality of Kaloyanovo has an important transport-geographic position. The main thoroughfares, which connect the north and south part of Bulgaria, respectively the countries from North-East Europe and Scandinavian countries with Turkey and the Near East pass through its territory. The railway road Plovdiv-Karlovo passes through six of the settlements and the railway road Plovdiv-Hisarya passes through other three settlements. Kaloyanovo is situated 24 km. from the regional center Plovdiv and 16 km. from "Trakia" highway, which connects Western Europe with Near East.

The selected municipality is good representative for analysing agricultural practices and making important points of preliminary knowledge about environmental structure. The extensive research of natural factors gives opportunities to increase positive sides of vegetable production. So it is essential to have profound analysis and necessary information before doing agricultural actions in some area.

MATERIALS AND METHODS

The base information about land management for making successful vegetable developing presents the present situation of land use and land management. The necessary data contents coordinated geographical borderlines of villages in the Municipality of Kaloyanovo, road systems, water resources, land using data, public or social property, population, classify-cation of cultivated terrains. All information is gathered from Municipality of Kaloyanovo and statistic data from different researches. Some of them are:

- Cadastre maps- The digital model formats are ZEM, CAD. Information source: the Geodesy, Cartography and Cadastre Agency.
- Statistic data about population, land use, land category, road and water systems-source http://www.kaloianovo.org/.
- Soil characteristics- Information source: The Soil Resources Agency and the Insti-tute of Soil Science "Nikola Pushkarov".

All action related to spatial data as collecting new information, organize in groups, creating connection between them, logical links, correct and sufficient presentation and sharing can be realized by Geographical Information Systems, named GIS (Stefanova et al., 2014). Methodo-logy is based on using GIS platforms and application for analysing and presenting the results. All transformation actions of collected data are made by ArcGIS software and appro-priate spatial data filters. The results are thematic maps of collected information.

RESULTS AND DISCUSSIONS

Multi-criteria analysis is complex information from different branches. It includes geographic information, coordinated location of the studied area, climate, soil type, elevation, population, demographic data, statistic information, agro-ecological settings, land use and management, etc. All necessary information is collected and arranged carefully by different experts from various scientific branches. Information transformation from paper to spatial data and introduced into computer managed software is obligatory action for easily dealing with a huge amount of heterogeneous data. The core of this study is presenting the necessity of background agricultural analysis and making well-arranged future actions for more profits. Brief description of the study area in general is presented by thematic maps about some of the important agricultural points. Attributes of the study area has marked effects on the tradition and culture and in turn to the cultivation practices. The information would show the basic facts to be considered for the data analysis and interpretation of the results. Environmental factors and technics progress urge forward making consideration of the study area information and put it on prime importance. It includes information from the socioeconomic, demography, land use and vivid dimensions. All data is collected from different sources, geographic transformed and coordinated and introduced into GIS software. The spatial data can be easily manipulated, manual added, clearly presented and used for future predictions. GIS application allows additions in every time and enriching the past and present data with new information. So GIS tools allow making multi-criteria analysis and combined various data from different branches. The studied area is situated in South Bulgaria and it is a part of Plovdiv region. This is the most developed agricultural region for vegetable production (Arnaudova et al., 2014). Today, Municipality of Kaloyanovo faces a number of various environmental and ecological innovations and yield increased challenges. The development and implementation of an environmental action plan for the valleys are associated with the strategy position and closely connection with another neighbour agricultural areas. The picture above presents the Municipality of Kaloyanovo position into Bulgaria map (Figure 1).

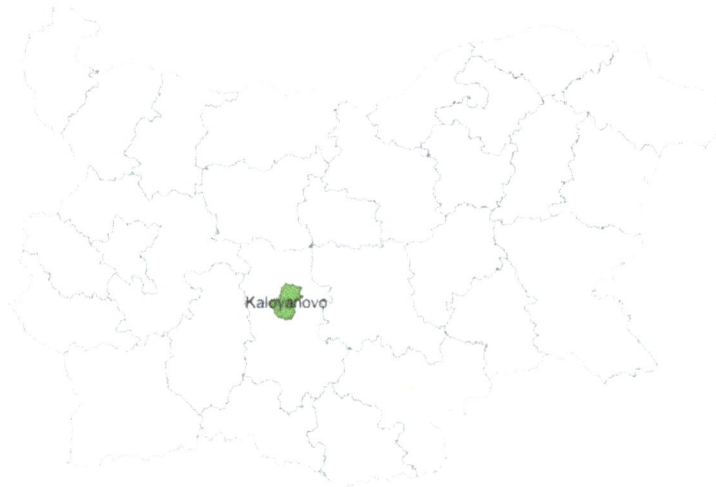

Figure 1: Bulgaria map and Municipality of Kaloyanovo (original)

In the municipality are situated 14 villages with one center city, named Kaloyanovo. The bigger village is named Razhevo Konare with area of 46.01sq.km and the smallest one is Glavatarwith area of6.23 sq.km. Next map (Figure 2) presents the position of 14 villages in Municipality Kaloyanovo. The relief is plane to hilly and average elevation is 250 m. The climate is transcontinental, characterized with an open winter and a hot summer. These factors are essential and favorable for vegetable developing.

Next maps (Figures 3 and 4) present information about land use and soil distribution in the studied area. Dominant soil types are sandy-clay and loamy, presented in the next map (Figure 3). These soil characteristics are useful for cultivation of various vegetable plants- tomatoes, papers, cabbages, carrots, etc.

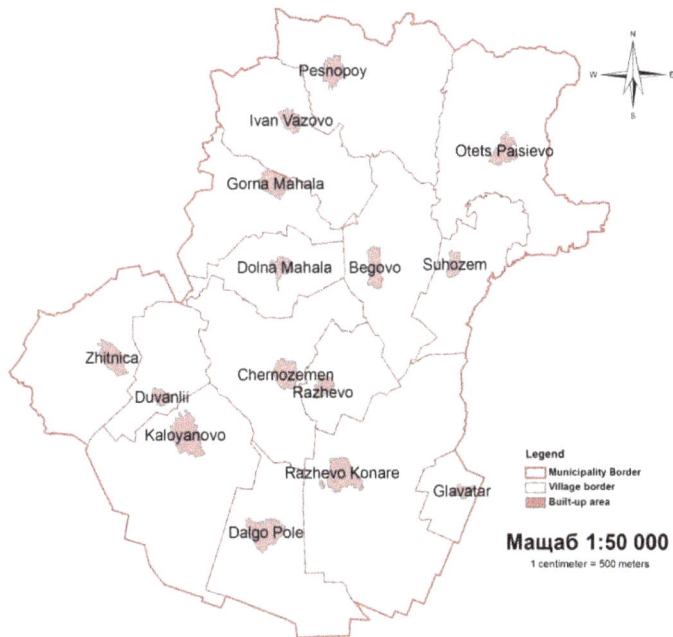

Figure 2: Map of the studied area- Municipality of Kaloyanovo (original)

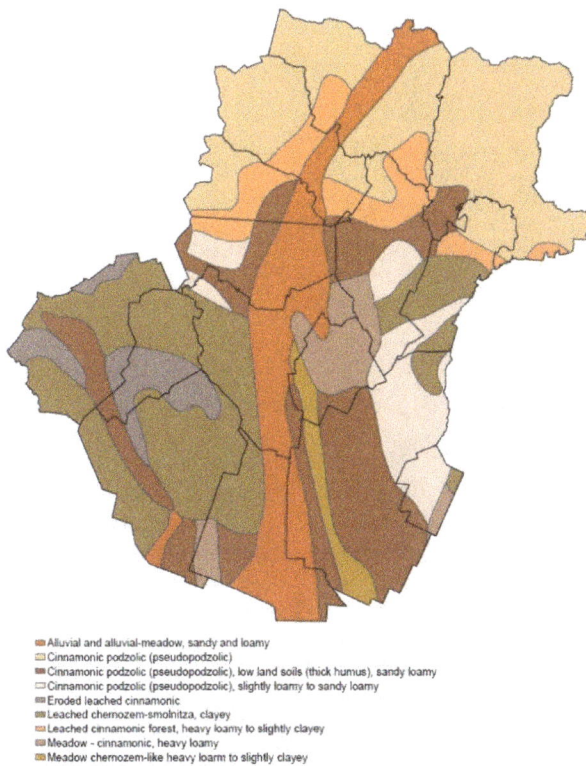

Figure 3: Map of Soil type in the studied area- Municipality of Kaloyanovo (original)

Good environmental factors advantage good agricultural actions. Above 50% of the land area is cultivated. Only the latest north terrains are covert with forest. More over 75% of the cultivated land is covered by corn fields, vegetable sorts and fruit plants. Forest area is 14,5%, almost 3.99% of urban area, 4.16%

water resources and another 2.35% for road systems, useful fossils, depots, scraps, etc. More agricultural land allows cultivating and developing of more different vegetable plants and opportunity of increasingly yields. All this information is presented by the next map (Figure 4).

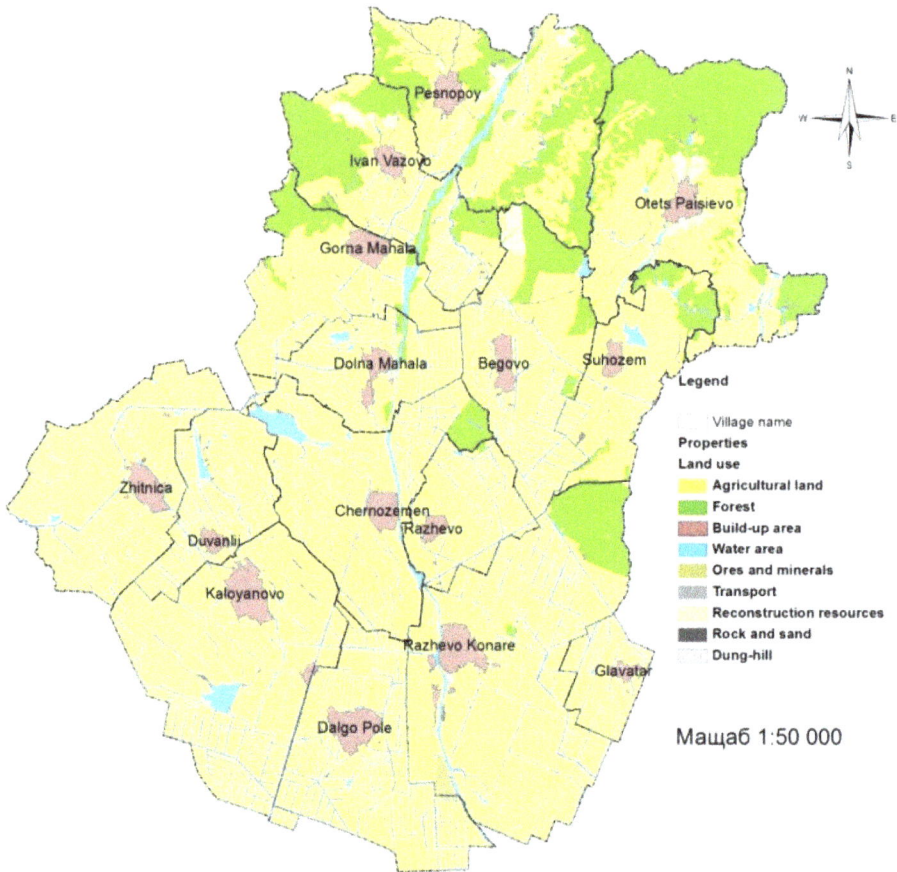

Figure 4: Land use map of Municipality Kaloyanovo (original)

According to the property of land in Municipality of Kaloyanovo (Figure 5) almost all cultivated area is private – 60.22%. State property is 16.8%, municipal area is 18.66% and built-up area 4.32%. So the most appropriate area for agricultural actions is private terrains. Private section is more developing and can be in funds by its own profits. So it is important for vegetable farming systems to be subsidized and to rely on its own production.

Categorization of cultivated land area (Figure 6) is made by 10 categories of comparison.

This method of land classification is the most spread in Bulgaria and it is consistent with climate conditions, relief, soil characteristics, plant requirements, etc. The classification is based on number from 2 to 10. Land category is assessment of land characteristics. The lowest number shows the best land for cultivation. Dominant land category is between 2 and 4, so it is very good for agricultural actions. Categorization is presented on the next map and categories are shown by different colors.

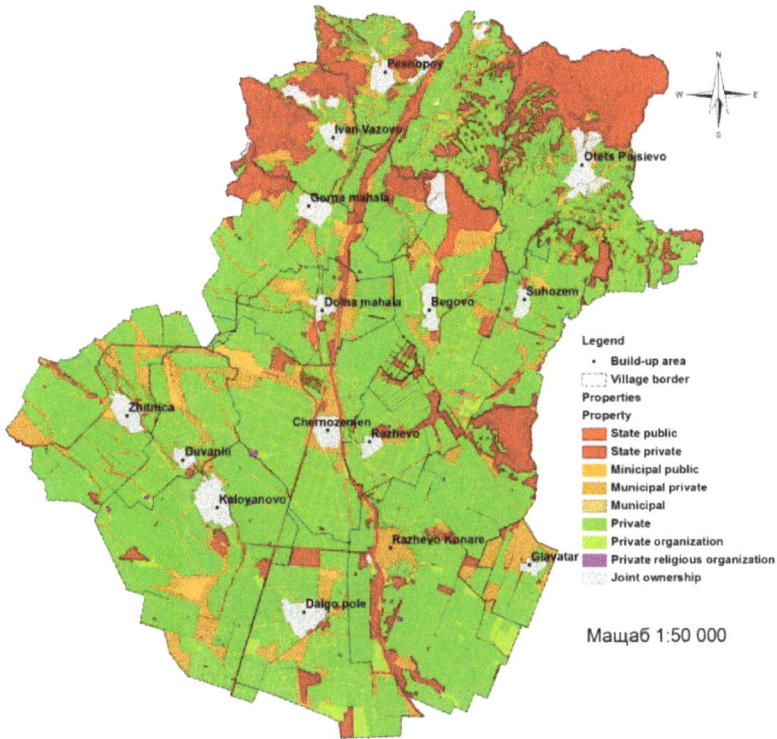

Figure 5: Property land map of the studied area (original)

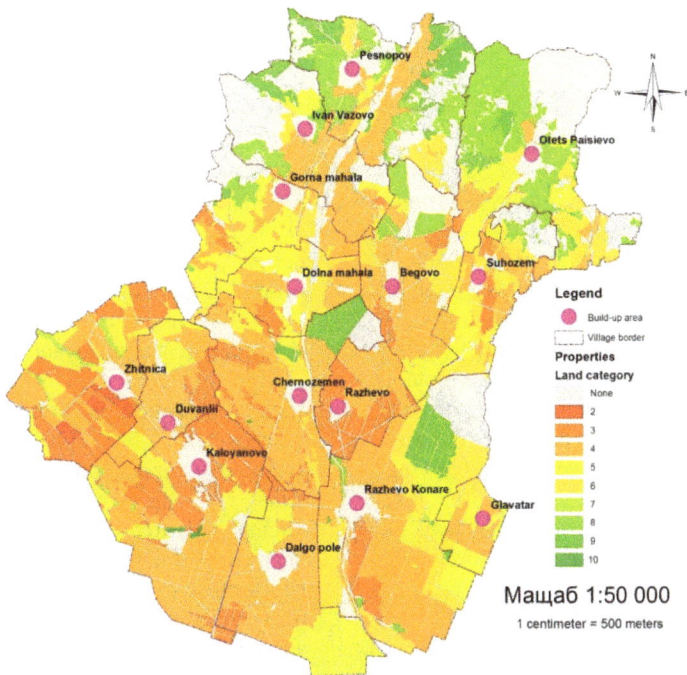

Figure 6: Map of Land category in Municipality Kaloyanovo (original)

Transport system (Figure 7) is amount to 152 km. It mainly contents roads from class 2 with length 24 km, class 3 with length 21 km and class 4 with length 62 km. During the studied area is situated railway with length 45 km. Municipality has important transported role of connecting North with South Bulgaria. Road information has influence on fast transportation of agricultural production to the market systems. Good structured roads help for connecting to the nearest factories and vegetable shops.

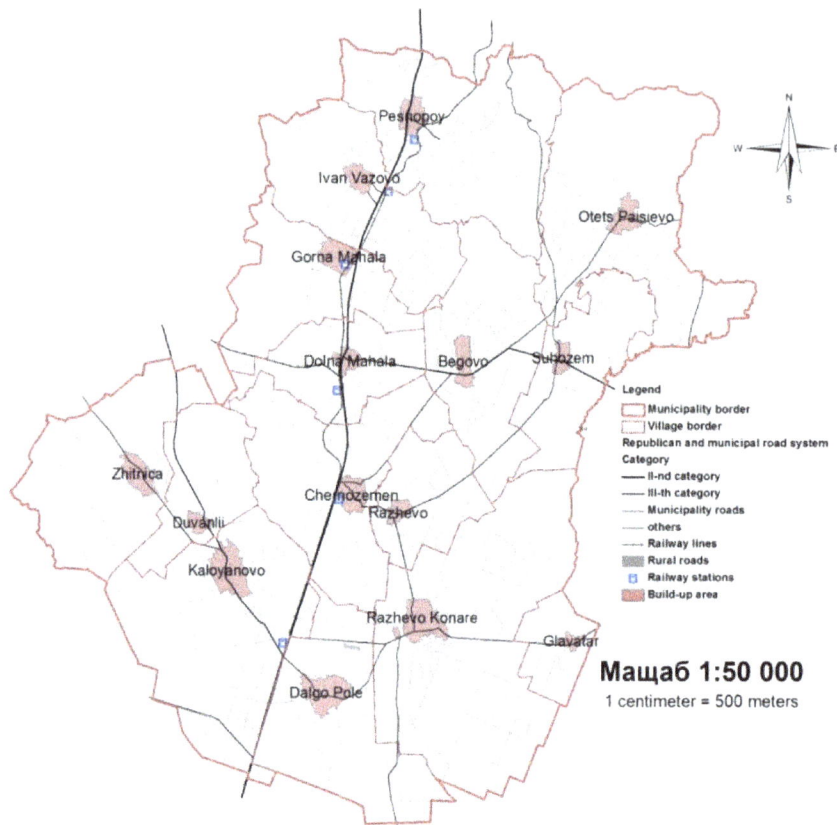

Figure 7: Map of road system (original)

Water resources (Figure 8) covered 14 500 ha area. Water system is presented by irrigation systems, rivers and canals. Length of water systems is 116 km. Next thematic map illustrates the situation of present irrigation systems and water resources in the Municipality of Kaloyanovo. Water resources are very substantial for vegetable growing. Analysed data face on good built and supported irrigation systems, which is permanently used and upgraded for agricultural practices.

Demography situation is presented on the next map (Figure 9). By the last census of the population in the studied area, organized during 2011 year, the population is 11 879 people. Dynamic is shown about the period 1934-2007 years. The most developed area is municipal center Kaloyanovo (Figure 10), so it is center city. In comparison with past years, population is decreased from small villages and move to the cities. The new technology tempts young people to the bigger cities, so the smaller villages become uninhabited areas. These results will have negative reflection on future agricultural development. It can be improved by implanting new facilities and more modern agricultural machines.

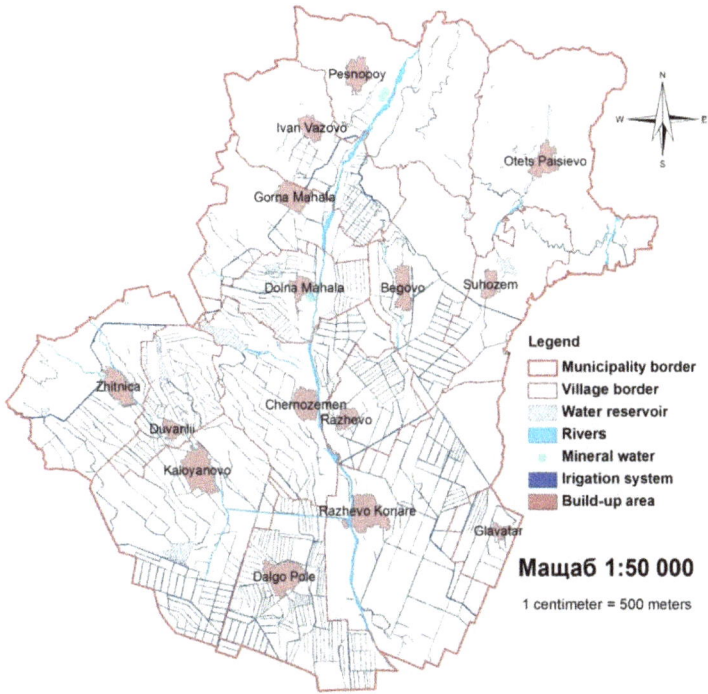

Figure 8: Map of water resources (original)

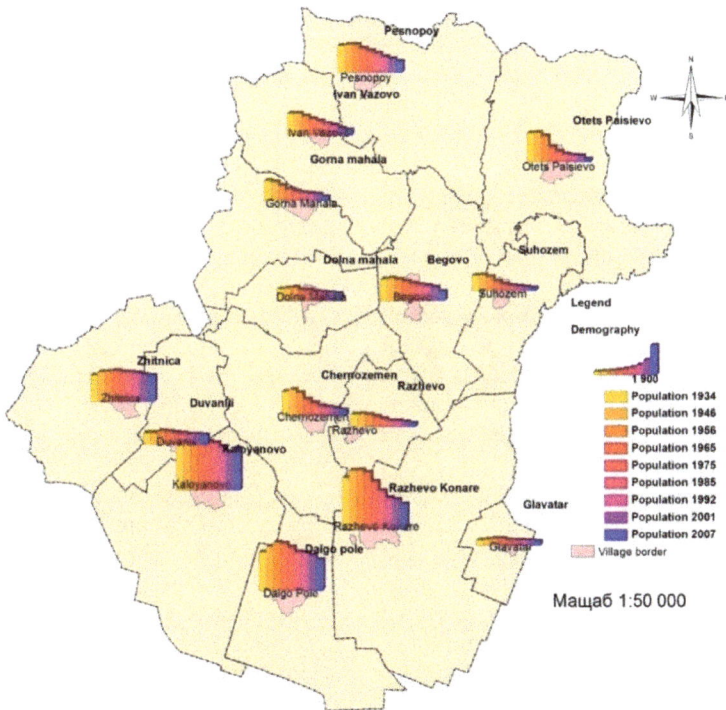

Figure 9: Demography map of studied area (original)

The main objective of the land evaluation is the prediction of the inherent capacity of a land unit to support a specific land use for a long period of time without deterioration, in order to minimize the socio-economic and environmental costs (De La Rosa 2000). Finding suitable land area for demanding agriculture crops is the need of present day farming system. So it is essential to know in advance if one area is good for agricultural practice or not. Background study is important part of farming management and it is the base level of expert knowledge.

All presented information is linked to the latest years. Spatial data transformation and implementation it in GIS show an easily way to analyze present information about nature and environmental resources. Agricultural development depends on environmental facilities. For good success and minimized losses, farmers have to do advanced research about environmental factors and nature resources. Results from the study have to be suitable for all plans requirements.

Figure 10: View from the studied area (http://www.kaloianovo.org)

CONCLUSIONS

Agriculture is a sector influenced by different parameters: nature resources, land use and management, water and road systems, population, plant requirements, etc.

All information is combined and presented by GIS tools. The data related to spatial information and boundary are calculated with the scale of 1:50000. Resulted thematic maps present the importance of environmental and human practices.

Municipality of Kaloyanovo is good area for agricultural development. All natural factors-soil distribution, water resources and land category will have positive effects on crops development. Good arranged road system and private land property increase product transportation to the vegetable markets and factories.

All nature factors have positive effects on distribution of land use - mainly cultivated areas.

As a negative factor is a result about decreasingly population in the smaller villages. It may be change by using more modern techniques.

Cultivation is difficult decision making and based on a huge factors. So advanced multi-criteria analysis is the method of mixing necessary information. GIS being one of the powerful tools, efficacy of the evaluation process will be maintained (Baniya, 2008). Non spatial parameters can also be analysed in the spatial basis to help making decision process easier. It is very essential to understand land capacity to support appropriate plants cultivation. So farmers are advised to use research for vegetable cultivating, according to nature and environmental potentiality. Site specific classification in order of suitability is main interest of vegetable growers for further profitable vegetable crop development.

ACKNOWLEDGEMENTS

There port and participations on the congress was financial supported by Project 05-14 of Research Centre in Agricultural University, Plovdiv (Bulgaria).

REFERENCES

Arnaudova Zh., Stefanova V., Haytova D., Bileva T., 2014. Gis based analysis of tomato and pepper growing regions in Bulgaria, Turkish journal of agricultural and natural sciences Balkan agriculture congress special issue: 1, www.turkjans.com

Baniya N., 2008. Land suitability evaluation using GIS for vegetable crops in Kathmandu Valley, Nepal. p. 244, Berlin (alsoavailableat: http://edoc.hu-berlin.de/dissertationen/baniya-nabarath-2008-10-13/PDF/baniya.pdf)

Christova E., Ilieva D., 2013. Productions of vegetables and fruits – potential for increasing employment in rural Rousse district, Proceedings of University of Ruse, vol.52,book 1.1,pp 122-125. (Bulgarian)

Davidson, D.A., 1992. The evaluation of land resources, Stirling University published in USA with John Wiley, NewYork.

De La Rosa D., 2000. MicroLEIS 2000: Conceptual Framework: Instituto de RecursosNaturalesy Agrobiologia, CSIC, Avda. Reina Mercedes 10, 41010 Sevilla, Spain: 267p. 212

Fresco L. O., Huizing H., Keulen H., Luning H.A., &Schipper R.A. 1994. Land Evaluation and Farming Systems Analysis for Land Use Planning. FAO WorkingDocument.

Nikolova M., 2013, Condition and challenges for Bulgarian agriculture after accession to the EU, Proceedings of University of Ruse, vol.52, book 5.1, pp 209-214. (Bulgarian)

Stefanova V., Arnaudova Zh., Haytova D., Bileva T., 2014. Multi-criteria evaluation for sustainable horticulture, Turkish journal of agricultural and natural sciences Balkan agriculture congress special issue: 2, p.1694-1701, www.turkjans.com

Stoeva T., 2013; "Economic effectiveness of vegetable production in Plovdiv region", Thesis, Agricultural University of Plovdiv, Bulgaria (Bulgarian)

Toskov G. , 2013, Main factors and tendencies analysis for competitivenes sincreasing of the vegetable production, Scientific Works of the Agricultural University, USAID, 1988. The transition to sustainable agriculture: an agendafor A.I.D: Committee for Agr. Sust. For Developing Countries, Washington D.C.

27

MAINTAINING OF THE QUALITY AFTER HARVESTING OF THE WHITE CABBAGE, DEPENDING ON NITROGEN FERTILISER DOSES APPLIED TO THE CROP

Constanța ALEXE[1], Marian VINTILĂ[1], Gheorghe LĂMUREANU[2], Adrian CHIRA[3], Lenuța CHIRA[3]

[1]Research and Development Institute for Processing and Marketing of the Horticultural Products – Bucharest- HORTING, No. 1A, Intrarea Binelui Street, District 4, 042159, Bucharest, Romania, E-mail: ihorting@yahoo.com
[2]Research Station for Fruit Growing (R.S.F.G) Constanta, No.1, Pepinierei Street, 907300, Commune Valu lui Traian, Romania, E-mail: scpp_constanta@hotmail.com
[3]University of Agronomic Sciences and Veterinary Medicine Bucharest, No. 59, Mărăşti Blvd, District 1, 011464, Bucharest, Romania, E-mail: lenutachira@yahoo.com

Corresponding author email: constanta_alexe@yahoo.com

Abstract

The purpose of this paper is to establish the influence of nitrogen fertilization of the cabbage crop on its preserving capacity. The paper presents the results obtained in the period 2014-2015 on quality maintaining of the white cabbage after harvesting. There were studied two varieties of cabbage: Gredana - summer, Danish variety and Buzoiana, autumn, Romanian variety. The culture was fertilized with nitrogen in the following doses: 100; 200; 300 and 400 kg active substance N/ha. Cabbage was kept in cold conditions (temperature = 0-2°C, relative humidity = 85-90%) for a period of 50 days - summer variety or 110 days - autumn variety. After storage determinations were performed on the weight losses (expressed by evaporate-transpiration), the losses by conditioning (resulted by removing yellowed leaves, of diseased ones and of the spine) and the main biochemical components. Researches show that the influence of nitrogen fertilization on maintaining of the quality of white cabbage during the preservation period is negative only at high doses (over 200 kg active substance N/ha). In case of dosages of 300 and 400 kg active substance N/ha losses were recorded of up to over 30% after 110 days of preservation of autumn cabbage. Therefore, in the case of cabbage destined to preservation it is not recommended the use of fertilizers in dosages greater than 200 kg active substance N/ha. The value of the main biochemical components is not strongly influenced by the dose of fertilizer applied to the crop. However one can notice an increase in the protein substances and a slight decrease of the values of ascorbic acid, total sugars and cellulose, from variant control (unfertilized) to variant with 400 kg active substance N/ha.

Key words: storage capacity, losses, biochemical components

INTRODUCTION

The therapeutic virtues of cabbage - this so common aliment - are known since antiquity. Its qualities are undeniable; therefore cabbage can be used successfully in the prevention and treatment of a very large number of diseases, being a real natural pharmacy. White cabbage is rich especially in pro-vitamin A, vitamins C and E, vitamin B1, vitamin B2, vitamin PP and in fibers, elements that provides cells health, giving it valuable therapeutic properties. Cabbage is richer in vitamin C than oranges, has few calories and a lot of substances with antioxidant effect, contains large amounts of magnesium, potassium, calcium, iron, copper, phosphorus, sulfur. These features and many others recommend the cabbage as a natural remedy against a large number of diseases (Popoescu et Zavoianu, 2011; Bogoescu, 2015). It is an alkalizing, nutritional, energetical, remineralizing and tonic aliment, and it is preferable to be eaten raw, in order to keep intact its properties. The research conducted in the last 30 years (Salunkhe et al.,1985.) confirmed that a regular consumption of cabbage has a beneficial effect in the prevention of the colon cancer in particular, of stomach cancer, but also of lungs, esophageal and rectal cancer (researches conducted at

University of Minnesota and J. Hopkins from USA, in Greece, Israel, Japan, Norway). Storage capacity of the cabbage depends on the quality of raw material for storage, which is influenced by culture conditions and variety (Ciofu, 2011, Ciuciuc et Toma, 2007, Stoleru et al., 2011). The variety imprints to the plants a certain chemical, histological and cytological structure (Jelea et Jelea, 2007, Salunkhe et Kadam, 1998, Stoleru et al., 2011), detectable through different methods of analysis (Hura, 2006). As far as produce destined for long term storage are concerned, as it is the case for cabbage, respecting the technological links which influence the forming and maintaining of the quality and the storage capacity is of great importance because choosing the adequate type of cabbage for the storage spaces is of the essential importance when aiming to maintain the quality during storage. In order to achieve this goal the crops destined for storage must be kept under observation beginning with the growing periods, meaning the moment the crop is started.

The purpose of this paper is to evaluate the preserving storage capacity of summer white cabbage of the 'Gredana' variety and of autumn cabbage, 'Buzoiana' variety, depending on the doses of nitrogen fertilization during culture period.

MATERIALS AND METHODS

The researches were conducted during period 2014-2015, using summer and autumn white cabbage, obtained in a vegetable farm located in an area of the Romanian seaside.

The trial was organized as a bi-factorial experience, with following experimental factors:
- A - variety
    a1 - Gredana (summer variety)
    a2 - Buzoiana (autumn variety)
- B - fertilization level
    b1 – control (unfertilized)
    b2 – 100 kg active substance N/ha
    b3 – 200 kg active substance N/ha.
    b4 – 300 kg active substance N/ha.
       (for summer variety, only)
    b5 – 500 kg active substance N/ha.
       (for autumn variety, only)
The storage was effectuated in refrigeration conditions (temperature = 0-2°C; air relative

humidity = 85-90%) for a period of 50 days (summer cabbage) and 105 days (autumn cabbage), thereupon the following determinations were effectuated:
- weight losses, resulted by evaporate - transpiration;
- losses by conditioning, resulted by removing yellowed leaves, of diseased ones and of the spine.
- identification of pathogens that caused the rot of the cabbage;
- main biochemical components (soluble solids, soluble sugars, titratable acidity, ascorbic acid, protein substances, cellulose).
The methods for determining the biochemical components were the following:
- the refractometric method, using the ABBE refractometer in order to determine the content of soluble dry substance;
- the Bertrand titrimetric method, for the determination of the content of soluble carbohydrates;
- the titrimetric method, for the determination of the titratable acidity;
- the spectofotometric method, for the determination of the ascorbic acid
- the gravimetric method, for the determination of the cellulose;
- Kjeldahl method, for the determination of the proteins.
During storage the hydro-thermal factors in the storage room were verified on a daily basis in order to ensure the respecting of the optimal conditions for the maintaining of the quality. Also, appreciations were made concerning the cabbage' capacity to maintain their quality during storage, as well as the possible occurrence and development of various specific diseases.

RESULTS AND DISCUSSIONS

1. Level of losses
Data presented in Table 1, on the preservation for short time of the summer cabbage, show that between the three graduations of fertilization with nitrogen that were studied, after 50 days of cold storage, there were no pronounced differences of the losses. It finds that weight losses recorded values very close to the three variants: from 4.8% in the case of fertilization b1 variant (unfertilized) to 5.3% in the case variant b3, with the dose of

fertilization of 200 kg active substance N/ha.

Table 1. Loses recorded by summer white cabbage after 50 days of cold storage

Variant	Fertilization level ((kg a.s. N/ha.)	Losses (%)		
		quantitative losses	qualitative depreciation	total
b1	unfertilized	4.8	5.0	9.8
b2	N 100	5.2	5.1	10.3
b3	N 200	5.3	5.5	10.8

Because of during storage period have not reported disease attacks, the conditioning consisted only in the removing of 1-2 yellowed leaves, which made that the conditioning losses be small (between 5.0% in the case of variant b1 and 5.5% in the case of variant b3). Total losses recorded values between 9.8% at the control variant and 10.8% at b3 variant.

Based on these results it can be considered that the fertilization with nitrogen in amounts up to 200 kg don't influence negatively the preservation capacity of summer cabbage, 'Gredana' variety, if it is storage a period of 55 days.

Analyzing the data presented in Table 2, on the preservation of long time of autumn cabbage, variety 'Buzoiana', it finds that after 110 days of refrigerated storing, weight losses increase proportionally to the dose of nitrogen: from 6.0% in the case of control, up to 10.8% in the case of variant b5, fertilized with 400 kg active substance N/ha.

Qualitative depreciation also increased proportionally with the dose of nitrogen applied to the crop. They are between 17.6% in the case of the control variant and 20.7% in the case of variant b5, but the differences are not significant compared to control only at variant b5.

Unlike summer cabbage, kept a short period, were there was no attack disease, at keeping for a long period of the autumn cabbage, there appeared various diseases and the conditioning losses resulted largely by removing rotting leaves.

Among the pathogens that caused the rot of outer leaves can mention: *Botrytis cinerea, Sclerotinia sclerotiorum, Alternaria brassicae* and bacteria of the genus *Pseudomonas*. The attack was emphasized at the cabbage of b4 and

b5 variants. In addition, at variant b5 were found brownings of the inner leaves oh the heads.

Table 2. Loses recorded by autumn white cabbage after 110 days of cold storage

Variant	Fertilization level (kg a. s. N/ha.)	Losses (%)		
		quantitative losses	qualitative depreciation	total
b1-control	-	6.0	17.6	23.6
b2	N 100	6.9	17.8	24.7
b3	N 200	7.8*	18.7	26.5
b4	N 300	8.3**	19.1	27.4*
b5	N 400	10.8***	20.7*	31.5***
		DL 5%=1.70	DL 5%=2.33	DL 5%= 3.69
		DL 1%= 2.48	DL 1%=3.39	DL 1%=5.36
		DL 0.1%=3.72	DL 0,1%=5.09	DL 0,1%=8.05

Total losses ranged from 23.6% at the control variant to 31.5% at variant b5, differences between these variant being very significant. In case of variant b4 differences were significant, compared to the control.

From the data presented up to now, it results that the effect of nitrogen fertilization on the storage capacity of the cabbage is negative only at high doses (over 200 kg a.s. N/ha). Until this dose there have not been reported the negative effect neither summer cabbage nor autumn cabbage, for which we can consider as limit to which we can apply nitrogen fertilizer without affecting the storage capacity of the cabbage is 200 kg a.s. N/ha.

2. Level of the biochemical components.

The analysis of the main biochemical components of the autumn cabbage (Table 3) reveals the fact that between the variants of nitrogen fertilization does not occur essential differences from this point of view, except protein substances that, from the value of 0.64% seen in variant b1, increase to 1.12% at variant b5.

It is observe a slight decrease in values of the ascorbic acid, total sugars and cellulose, from variant b1 to variant b5.

It also noted that during the storage of cabbage, the quantities of the biochemical components did not modify noticeable.

Table 3. The main biochemical components of autumn cabbage at the beginning and after the cold storage

Specification	M.U.	Moment of determ.	Variant				
			b1	b2	b3	b4	b5
Dry soluble substance	%	I*	6.4	6.3	6.0	6.0	6.0
		II**	6.5	6.4	6.1	6.1	6.1
Ascorbic acid	mg/100g	I	35.99	35.11	34.73	34.50	32.64
		II	35.15	34.80	33.45	32.30	30.40
Titratable acidity	%	I	0.15	0.15	0.15	0.15	0.14
		II	0.16	0.16	0.15	0.16	0.15
Total carbohydrates	%	I	3.68	3.58	3.68	3.42	2.94
		II	3.23	3.48	2.94	2.75	2.80
Proteins (Nx 6.25%)	%	I	0.64	0.71	0.76	0.80	1.00
		II	0.80	0.87	0.88	1.05	1.12
Cellulose	%	I	0.73	0.73	0.72	0.63	0.61
		II	1.17	0,96	0.85	0.78	0.88

* at harvest
** after 110 days of cold storage

This is explained by the fact that the cabbage was subjected to low temperature and so by the slowing of the metabolic process. Very small increases observed after storage in dry matter values, titratable acidity and cellulose, but content in ascorbic acid and total sugars decrease.

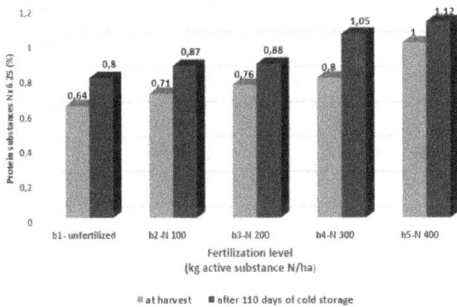

Figure 1. The proteins content of the autumn cabbage, fertilized with different nitrogen doses

CONCLUSIONS

Fertilization of the white summer and autumn cabbage with nitrogen fertilizer up to 200 kg active substances/ha not affect the preservation quality of the cabbage.
After 50 days of storage of the summer cabbage and after 110 days of storage of the autumn cabbage the total losses are insignificant compared to the control.
At higher doses of nitrogen fertilizers, 300 and 400 kg active substances N/ha, the capacity for storage of autumn cabbage was affected. After 110 days of storage total losses were over 30%.
Therefore, for the cabbage destined for preserving for long time is not recommended fertilizer doses greater than 200 kg N/ ha.

The value of the main biochemical components of the autumn cabbage is not strongly influenced by the dose of fertilizer applied to the crop.
In the case of different fertilization level, there are no essential differences in the values of the main biochemical components, beside the proteins. There was an increase in values at protein substances (0.64%-b1 variant, 1.00%-b5 variant) and a slight decrease in values of ascorbic acid, total sugars and cellulose, from control variant to variant b5 (400 kg active substances N/ha).
Because of cold storage, intensity of the metabolic processes decreased, for which, during storage, the biochemical component did not suffer major changes.

REFERENCES

Bogoescu M., 2014. Valorificarea verzei. Ed. Cermaprint, Bucuresti, 10-18.
Ciofu R., 2011. Tratat de legumicultura Ed. CERES, 688-718.
Ciuciuc E., Toma V., 2007. Comportarea unor cultivare de varza alba in conditiile solurilor nisipoase. Lucrări științifice CCDCPN Dabuleni, vol.XVI, 49-56.
Jelea S.G., Jelea M., 2007. Citologie. Histologie. Embriologie. Ed. Universităţiii de Nord, Baia Mare
Hura C., 2006 . Metode de analiză pentru produse alimentare, ghid de laborator. Editura Cermi, Iaşi
Popoescu V., Zavoianu R., 2011. Cultura legumelor din grupa verzei. Ed. M.A.S.T., 26-48
Salunkhe D.K., Kadam S.S., 1998. Handbook of Vegetable Science and Technology: Production, Compostion, Storage and Processing. CRC Press, 299-327.
Salunkhe D. K., Kadam S.S., Chavan J.K., 1985. Postharvest Biotechnology Of Food Legumes Hardcover, 115-129.
Stoleru C. M., Stan N., Stan C., 2011. Influenţa unor factori tehnologici asupra calităţii producţiei la varza cultivată în sistem ecologic. Lucrări Ştiinţifice - vol. 51, seria Agronomie, 369-374.

28

EFFECT OF SOME "BIO-INSECTICIDES" USED AGAINST TWO SPOTTED SPIDER MITES (*TETRANYCHUS URTICAE* KOCH.) IN THE CUCUMBERS CROP UNDER PLASTIC TUNNEL CONDITIONS

Ana Emilia CENUŞĂ[1], Gabriela ŞOVĂREL[1], Marcel COSTACHE[1], Elena BRATU[1], Marius VELEA[2]

[1]Research and Development Institute for Vegetables and Flower Growing Calea Bucureşti 22, Vidra, Ilfov, Romania

[2]Holland Farming Agro SRL, Bucureşti, Drumul Osiei 74, sect. 6
Corresponding author e-mail: bratu.bpe@icdlfvidra.ro

Abstract

Experiment was performed in the summer of 2015, in a cucumbers crop under high plastic tunnel conditions in order to determine the efficacy of some bio-insecticides in the control of two spotted spider mites (Tetranychus urticae) attack. Bio-products Oleorgan - 0.3% (saponified neem oil 40%), Konflic - 0.3% (Quassia amara 50%+saponified extract of different oils 50%), Canelys – 0.3% (cinnamon extract 70%), Kabon – 0.3% (potassium soap from vegetable oils 50%), Zicara – 0.15% (citrus peel extract and essential oils 70%), Lasser 240 SC – 0.05% (spinosad 24%) and Vertimec 1.8 EC – 0.08% (abamectine 1.8%) were applied single or mixed, repetitive or alternated in different variants. Were performed 6 applications. Intervals between treatments decreased from 7 to 3-4 days, depending of pest infestation level. Percent of attacked area/leaf and attack/plant established by visual estimation, and the harvest/plant were registered at the end of crop vegetation cycle. In addition, for a better appreciation of obtaining results were calculated the average attack and production increasing. Analyzing of the all parameters take into account showed that the most relevant of them proved to be the average attack (noticed as "general attack"), harvest/plant and harvest increasing. Yield data were statistical assured. Based on these parameters, best results (6.1%, 3.914 kg, and 224.8%, resp.) were obtained with a combination of spinosad 24% and abamectine 1.8% applied alternate. Appropriate values (6.4%, 3.076 kg, and 176.7%, resp.) were registered with product Oleorgan based on saponified neem oil 40%. Relatively good results (17.7%, 2.504 kg, and 143.8%, resp.) have been obtained by applying a mixture containing Oleorgan + Kabon + Canelys (0.3% each) alternately with another mixture: Konflic + Canelys + Zicara (0.3%; 0.3%; 0.15%). In the cases of spinosad 24% and untreated check, the three parameters take into account presented values of 75 and 92.5% resp., 2.248 and 1.741 kg, resp., and 130.8% (untreated check was reference). The study revealed that the saponified neem oil 40% applied alone had lower but appropriate performances to those of spinosad and abamectine alternately applied, proving a high acaricidal action, which promote it to be used in IPM and organic farming practices.

Key words: *bio-insecticides, plant extracts, plant oils, efficacy, cucumber, mites*

INTRODUCTION

The use of bio-insecticides based on plant extracts or metabolites of the various categories of bodies has experienced a revival in the past decade because of its positive impact on public and enviromental health. Plant species whose insecticidal action was known long time ago came to the attention of scientists, along with the start of investigations for finding new botanical source-species (Grdiša and Gršić, 2013). Despite their beneficial effect on the quality of the environment, bio-pesticides in general have several limiting characteristics (quick degradation in sunlight, air and moisture) that make them less agreed by large farmers communities, being recognized and used mainly in organic farming from developed countries. They have initiated extensive researches to identify their own botanical source-species and the formulation of their own bio-insecticides (Khater, 2012).

The present work aimed to establish the action spectrum and efficacy of some comercial bio-insecticides applied alone or mixed, repetitive or alternatively to control populations of two spotted spider mites (*Tetranychus urticae*) in a cucumber crop under high plastic tunnel.

MATERIALS AND METHODS

Experiment was performed in a cucumber crop (cultivar „Mirabelle F1") under high plastic tunnel, natural infested with two spotted spider mites (*Tetranychus urticae* Koch.), during summer season, 2015 in Vidra-Ilfov, Romania. Tested bio-insecticides were: Oleorgan (saponified neem oil 40%), Konflic (*Quassia amara* 50% + saponified extract of different oils 50%), Canelys (*cinnamon* extract 70%), Kabon (potassium soap from vegetable oils 50%), Zicara (*citrus* peel extract and essential oils 70%), Lasser 240 SC (*spinosad* 24%) and Vertimec 1.8 EC (*abamectine* 1.8%).

Each of treatment variant was organized in three replicates (20 plants/replication), liniar arranged. The treatment variants were:

1. Mixed bio-insecticides (Oleorgan 0.3% + Kabon 0.3% + Canelys 0.3% mixed and alternately applied with Konflic 0.3% + Canelys 0.3% + Zicara 0.15% mixed);
2. Laser 240 SC 0.05% alone, alternately applied with Vertimec 1.8 EC, 0.08% alone;
3. Oleorgan 0.3% alone, repetitive applied;
4. Laser 240 SC 0.05% alone, repetitive applied;
5. Untreated check.

The bio-insecticides water solutions were applied by spraying of plant leaves, using a manual pump.

It were applied 6 treatments during the vegetation season, intervals between applications decreasing from 7 days (at the beginning of pest infestation) to 3-4 days (in the top of pest development).

A single visual observation was performed 4 days after last treament by this way being estimated the percent of attacked area/leaf (sample size: 4 leaves/5 plants) and the percent of attacked area/plant (5 plants/replicate). Yield was registered also, during experimental period and statistical analyzed.

RESULTS AND DISCUSSIONS

Obtained results had confirmed the scientific researches (Martinez, 2001; Hummel and Kleeberg, 2001) showing the great efficacy of saponified neem oil 40% (Oleorgan) in the control of two spotted spider mites populations (fig. 1).

In this case had emphasized a compensatory effect between low value of the attack/leaf (6.2 %) and that easy greater of the attack/plant (6.5

%) which proves that attacked areas on leaves were small but the number of infested leaves on plant was greater (fig. 2).

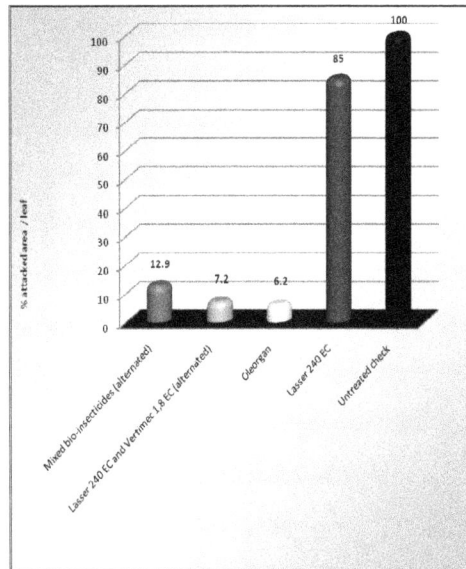

Figure 1. Effect of some "bio-insecticides" used in the control of mites on cucumbers crop under high plastic tunnel (attack/leaf)

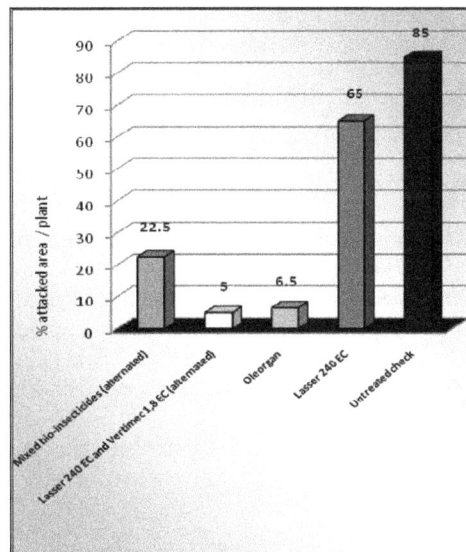

Figure 2. Effect of some "bio-insecticides" used in the control of mites on cucumbers crop under high plastic tunnel (attack/plant)

By alternating *spinosad* (Laser 240 SC) and *abamectine* (Vertimec 1.8 EC) applications the effect was opposite, the attacked areas on leaves being greater while the number of infested leaves on plant was lower.

For a balanced appreciation of the obtained results it was calculated the average values of the two registered parameters noticed as "general attack" (fig. 3).

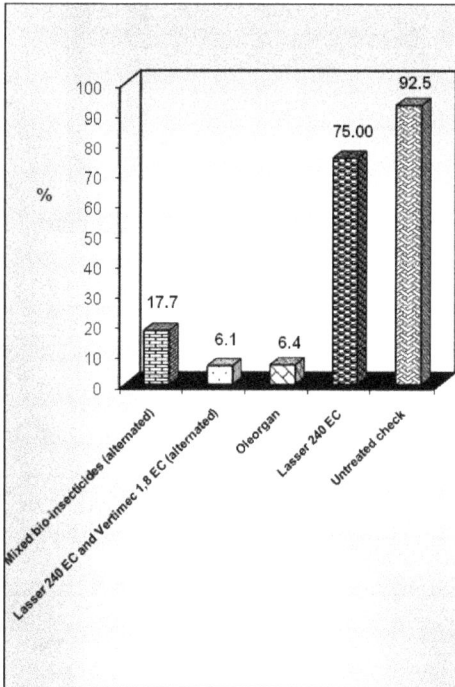

Figure 3. Effect of some "bio-insecticides" used in the control of mites on cucumbers crop under high plastic tunnel (general attack/plant)

They showed the similar effects of neem oil and alternatively treatments with *spinosad* and *abamectine* in the control of two spotted spider mites populations. Mixed bio-insecticides, alternatively applied gave good results also, by comparison with *spinosad* and untreated check. Despite of appropriate results obtained with neem oil and spinosad/abamectine alternate application, the differences between the yields of these two variants were bigger (fig. 4) suggesting a certain negative influence of oil compound on cucumbers plants. However, all the four treatment variants experimented gave better results than untreated check regarding harvest/plant. The biggest yield difference comparing to untreated check was obtained with *spinosad/abamectine* alternate applications followed by *neem* oil, mixed bio-

insecticides alternatively applied and *spinosad* repetitive applications (fig. 5).

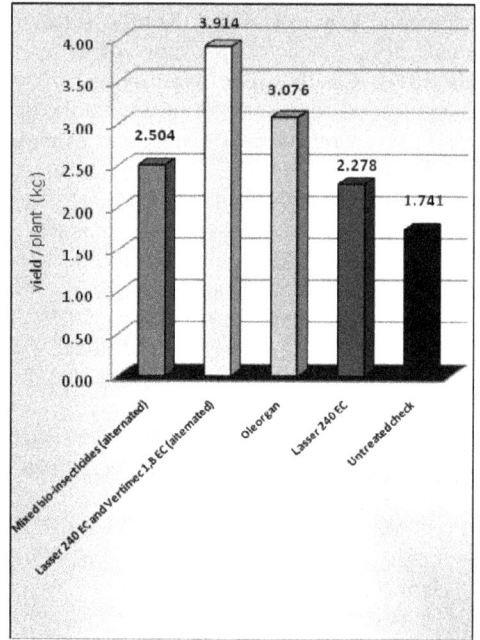

Figure 4. Harvest obtained in the cucumber crop treated with different „bio-insecticides"

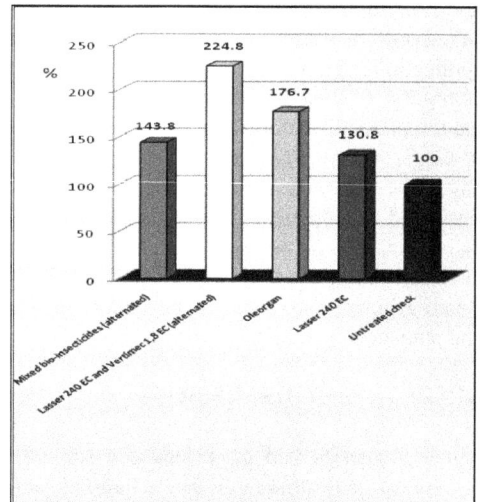

Figure 5. Harvest increasing at cucumbers crop treated with different „bio-insecticides"

Statistical interpretation of yield data were made by analysis of variance (table 1).

Table 1. Significance of yield differences between treated variants and untreated check

Variant	Yield (t/ha)	Yield (%)	Differences (t/ha)	Significance
1	31.73	143.8	+ 9.66	***
2	49.60	224.7	+ 27.53	***
3	39.00	176.7	+ 16.93	***
4	28.87	130.8	+ 6.80	***
5	22.07	100.0	-	-

LSD 5%	= 1.78
LSD 1%	= 2.59
LSD 0.1%	= 3.88

Differences between the experimental variants can be observed in the photos 1, 2, 3, 4 and 5.

Photo 1. Effect of mixed bio-insecticides in the control of red spider mites on cucumber plant (V1)

Photo 2. Effect of *spinosad* and *abamectin* alternate applied in the control of red spider mites on cucumber plants (V2)

Photo 3. Effect of saponified *neem* oil in the control of red spider mites on cucumber (V3).

Photo 4. Effect of *spinosad* in the control of red spider mites on cucumber plants (V4).

Photo 5. Effect of red spider mites attack on untreated cucumber plants (V5).

CONCLUSIONS

The most relevant parameters in this study proved to be general attack, harvest/plant and production increasing compared to untreated check. Based on them, it can appreciates that *spinosad* and *abamectine* applied alternately gave the best results (6.1 %; 3.914 kg; 224.8 %, respectively), being followed of *neem* oil alone (6.4 %; 3.076 kg; 176.7 %, resp.), mixed bio-insecticides alternate (17.7 %; 2.504 kg; 143.8 %, resp.) and *spinosad* alone (75 %; 2.278 kg; 130.8 %, resp.). At untreated check were registered the most poor results (92.5 %; 1.741 kg).

On the other hand, the relatively great decreasing of yield registered at *neem* oil by comparison with *spinosad* and *abamectine* alternately applied, it could suggest a slowly phytotoxic reaction to the oil compound that reduce the ability of plant respiration, but this should be studied in depth.

REFERENCES

Grdiša M., Gršić K., 2013. Botanical insecticides in plant protection. Agriculturae Conspectus Scientificus, vol.78, no. 2 (85-93).

Hanem F. K., 2012. Ecosmart Biorational Insecticides: Alternative Insect Control Strategies, Insecticides - Advances in Integrated Pest Management, Dr. Farzana Perveen (Ed.), ISBN: 978-953-307-780-2, InTech, DOI: 10.5772/27852.

Martinez S.S., 2001. Main pest which can potentially be controlled with techniques based on Neem products in Brazil. Practice Oriented Results on Used and Production of Plant Extracts and Pheromones in Integrated and Biological Pest Control; Abstracts of the 1. Workshop "Neem and Pheromones"; University of Uberaba, Brazil, March 29-30, 23-24.

Hummel E., Kleeberg H., 2001. Properties of NeemAzal[TM]-TS – experienced and possibilities in biological plant protection. Practice Oriented Results on Used and Production of Plant Extracts and Pheromones in Integrated and Biological Pest Control; Abstracts of the 1. Workshop "Neem and Pheromones"; University of Uberaba, Brazil, March 29-30, 10-22.

29

NEW GENOTYPES OF *LUFFA* SPP. OBTAINED AT V.R.D.S. BUZĂU

Costel VÎNĂTORU[1], Bianca ZAMFIR[1], Camelia BRATU[1], Liliana BĂDULESCU[2], Viorica LAGUNOVSCHI[2]

[1]Vegetable Research and Development Station Buzău, No. 23, Mesteacănului Street, zip code 120024, Buzău, Romania

[2]University of Agronomic Sciences and Veterinary Medicine of Bucharest, 59 Marasti Blvd, District 1, Bucharest, Romania

Corresponding author email: costel_vinatoru@yahoo.com

Abstract

Luffa cylindrica belongs to the Cucurbitaceae family, being an annual, herbaceous plant, multiplied by seeds. It is a well-defined genre, alongside with the species L. acutangula, L echinata,, L. graveolens, with their origin along with L. cylindrica in the tropics. And the species l. operculata, l. quinquefida and l. astorii are originating in the neo-tropical zone. Of these, these two species originating in India, L. cylindrica and L. acutangula are most common. They have been acclimatized for starters in India and America, then expanded and grown on a large scale for their immature fruits used as vegetables. With the passage of time, the areas of usage have diversified greatly. In Romania, the most known species is Luffa cylindrica, plant introduced in our country after the 1960s, at V.R.D.S. Buzau by dr. ing. Marcela Iosifescu. Though it has been successfully acclimatized in our country, especially in the protected spaces, promoting this culture was quite slow. At present, the spaces occupied by this species are pretty small, insignificant. After 1996, researches regarding the Luffa species were taken over time by the Laboratory of Improvement, aiming the acclimatization of new species and obtaining new creations with distinct biological phenotypical expression. Along with Luffa cylindrica, a special attention was given to the acclimation of new species, of which Luffa acutangula was successfully acclimated. The research continued with the crossing of L. acutangula x L. cylindrica species. By this crossing was obtained an F1 hybrid with intermediate sized fruit and high density of the fibre. From the segregation of the F1 hybrid were obtained in 6 new families, lineage with distinct features and numerous intermediate forms have been removed in the process of improvement. Varieties of Luffa acutangula have very large fruit, in green have recorded an average of 119 cm, unlike Luffa cylindrica, which recorded an average of 56 cm. F1 hybrid has an average length of fruit length of 65 cm. New varieties derived from the crossing of species ranged from L1 to L5 with 91 cm and 48 cm, the smallest value. The researches were completed with the obtaining of new genotypes Luffa cylindrica and L. acutangula with distinct features.

Key words: acclimation, hybrid, improvement, L. acutangula, L. cylindrica.

INTRODUCTION

Luffa species have exhibited a special interest for V.R.D.S. Buzău, which is also the first institution in the country where *Luffa cylindrica* was introduced in breeding programs by Ph.D. Engineer Iosifescu Marcela, after 1960. First researches have been channeled towards testing products to combat certain pathogens and diseases in *Cucurbitaceae*, knowing that *Luffa* spp. shows a remarkable resistance within this family.

The *Cucurbitaceae* or vine crop family is a distinct family without any close relatives and includes many important vegetables such as cucumber, melon, watermelon, squash, and gourds.

Plants within *Cucurbitaceae* consist of 95 genera (Kousik et al., 2015)

Is a well-defined genre, alongside the species *L. acutangula*, *L. echinata*, *L. graveolens*, which have their origins along with *L. cylindrica* in the tropics. The species *L. operculata*, *L. quinquefida* and *L. astorii* are originating in the Neotropical. Of these, two native species, from India, *L. cylindrica* and *L. acutangula* are most common. They have been acclimatized for starters in India and America, then expanded and grown on a large scale for their immature fruits used as vegetables. In time, the areas of *Luffa* spp. usage have diversified greatly.

Although stages of acclimatization and improvement of the species accounted for a great success in our institution, it did not occupy until now significant areas in culture and no Romanian variety was patented. After 1996, researches focusing on *Luffa* species were resumed in an intensive system with the aim of obtaining new genotypes with precise directions for use and also the acclimation of new species, with an emphasis on *L. acutangula.*

Recent researches confirm that *Luffa* species are plants with multiple uses. The fruit contains triterpenoid saponins: lucyosides A, B, C, D, E, F, G, H, I, J, K, L, M, ginsenosides Re, Rg1, etc. The leaf contains triterpenoid saponins: lucyin A, lucyosides G, N, O, P, Q, R, 21β-hydroxyoleanoic acid, 3-O-β-D glucopyranosyl - maslinic acid ginsenosides Re, Rg1; flavonoids: apigenin, etc. The seed contains polypeptides: luffins P1, S, luffacylin etc. (Partap et al., 2012). As an entomophilies species, preferred by pollinating insects, it presents numerous genotypes.

Rich morphological variability occurs in cultivated species of *Luffa* in different growing regions. (Prakash et al., 2013).

At the present, the interest for the species of *Luffa* grew considerably in our country, motivating the initiation of further researches presented in this paper.

MATERIALS AND METHODS

The researches started with the achievement and the enrichment of a germplasm collection for this species. Three species were taken in study: *Luffa cylindrica*, *Luffa acutangula* and *Luffa operculata*. Within the species *Luffa cylindrica* we managed to achieve a large number of genotypes but have been selected for study 3 genotypes with distinct characteristics and stable in descent, two of them obtained at the V.R.D.S. Buzău; one of them has white seed, the other one has black seed and the third one is from Bulgaria. Within the species of *Luffa acutangula,* researches started with the species acclimation, because so far these varieties were not cultivated in Romania. The germplasm collection was established with a total of five distinct genotypes, but one who has demonstrated adaptability and stability to

our soil and climatic conditions was the one originating from China, codenamed G2. Within the species *Luffa operculata,* until now, we haven't managed any genotype acclimation that presents adaptability and genetic stability.

The selection methods were the specific ones for cucurbits, and the stabilized families were subjected for hybridization followed by the segregation process. Special attention was paid to isolation areas due to its entomophilies degree. They were cultivated in different greenhouse compartments, to avoid contamination.

RESULTS AND DISCUSSIONS

Researches finalised with the achievement of a solid germplasm collection at this species.

From the *Luffa cylindrica* group, a new variety was achieved that presents distinct characteristics recommending it to be used as a vegetable sponge. The plant is vigorous, with a well developed root system that explores the deeper layers of soil.

In protected areas, the plant reaches the height of six up to eight meters. From the stem, eight-twelve main shoots with numerous secondary shoots develop, and have a capacity of dispersion of over six meters. The stem is vigorous, edged, slightly lignified at the base, with a medium diameter of 16 millimeters. (Fig. 1).

Figure 1. Stem

The plant has a rich foliar device, consisting of scattered leaves on shoots at a distance of 14-18 cm, with a leaf petiole ranging between 18-24 cm, and a diameter between 6-8 mm. Length of leaf varies between 20-30 cm and the width register values between 18-28 cm. (Fig.2.).

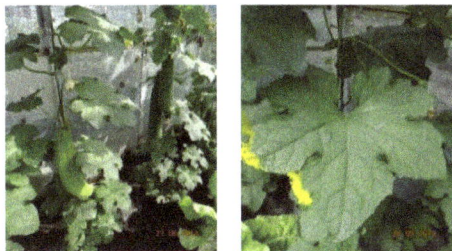
Figure 2. Plant and leaf detail

The leaf is composed of 5 lobes, the inferior ones being serrated. The shoots present tendrils to facilitate climbing on the trellising system (Fig.3).

Figure 3. Tendrils

The flowers are yellow, distinct on the plant; the male ones show a long petiole, which varies between 20-40 cm long, with an average diameter of 10 mm and the peduncle length of 1-1, 2 cm. The sepals are sharp and have an average length of 10 mm. The corolla diameter is of 8 cm, flowers are type-5, and number of flowers in florescence is 3-12. Female flowers are solitary and are distinguished by the presence of the miniature fruit at the base of the corolla. First make their appearance are male ones, that are far more numerous than females (Fig.4).

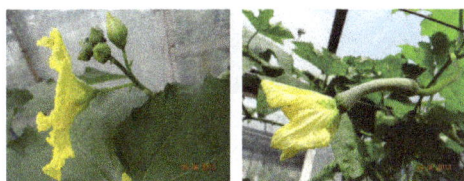
Figure 4. Male and female flower

The fruit is a green cylindrical ridged pepo, with the base slightly narrow than the apex. It has an inner dense network of cellulose fibre, a characteristic that gives it more quality in its use. The plant has a great capacity production, but in normal nutritional conditions it retains a number of 18-22 fruits per plant. (Fig. 5). The production can increase significantly if it interferes with special directing and topping of works for the unnecessary shoots and if is ensured an adequate level of nutrition per phenological phase.

Figure 5. Unripe fruit, dried fruit, sponge

Referring to the recent studied species, *L. acutangula*, the researches finalised with its acclimation, this being a first in our country, and with the achievement of a new variety with distinct characteristics. Regarding the biological specific features of this species, is similar with the *L. cylindrica,* bearing the mention that its fruit are likely to be consumed when fresh. Optimal size of fruit for consumption is of 20 - 40 cm. With this dimension, the fruit diameter at the base is of 6.5 cm, ant the apex diameter is of 7.2 cm. If are not harvested, the fruits lose their softness, spongy fibres appear inside and are unfit for human consumption. Such unharvested fruit can exceed 1 meter in length and as they mature, they turn from green to brown, lose significant weight and decrease greatly its size. (Fig. 6.)

Figure 6. Fruits at maturity consumption

Culture can be established through seedlings or direct sowing. In field conditions, the plant

behaves like a tardy plant, therefore it is recommended to be cultivated in the warmer areas of the country. Seedling production is carried out just like the rest of the cucurbits, being performed in alveolar palettes with 28 holes, in order to ensure an appropriate nutrition space. Sowing, for the production of seedlings is made in the first decade of February, and for the field in the first decade of March (Fig. 7).

Table.1. Main fruit characteristics

Variety name	Fruit length (cm)			Weight of the sponge (g)	Seed no./fruit	Seed weight/fruit (g)
	green	dried	sponge			
L. acutangula	119	98	93	41.7	534	64.1
L. cylindrica	56	47	44	26.3	312	37.4
Luffa CxA[1]	65	56	53	28.9	428	51.4
L1	91	82	77	57.8	444	53.3
L2	73	65	61	36.4	462	55.4
L3	63	52	58	22.2	512	61.5
L4	49	40	37	33.8	335	40.2
L5	48	36	32	5.5	279	33.5
L6	59	50	46	38.8	320	38.4

[1]Luffa C x A– Luffa cylindrica X Luffa acutangula

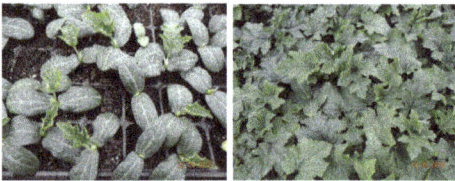

Figure 7. Seedlings

Planting in protected areas is carried out around April 1, while plantings in the field are made after 1 May. Seedling age should be 55-60 days. Where the establishment of culture is made by direct sowing, sowing in unheated protected areas is carried out after March 20 while in the field after April 20. (Fig.8).

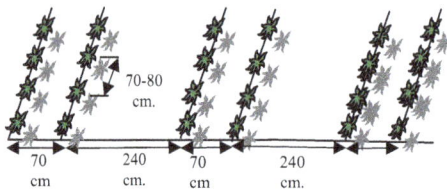

Figure 8. Crop on unshaped terrain establishment plan for protected areas

The species supports several technological variants. It can be established in equidistant rows or strips. It was found that the best results were obtained when the culture was established

in strips - 70 cm between rows and 240 cm between bands and between plants on the row 70-80 cm depending on the vigor of cultivar.

After the stabilization and the achievement of the two varieties of L. cylindrica and L. acutangula, the researches continued with the hybrid combination execution between the selected genotypes, aiming to obtain valuable hybrid creations and the enrichment of thr autochthonous genetic heritage.

Although the research conducted worldwide indicate that the crossing between the species L. cylindrica and L. acutangula presents sterility in descendents, the hybrid combination achieved at V.R.D.S. Buzău had demonstrated compatibility and manifested a visible pheno-menon of heterosis in F1, obtaining a hybrid with intermediate morphological features between the two species, a high yield pro-duction and a combined use destination. Young fruits can be used in nutrition and the mature ones are used as a vegetable sponge (Fig. 9).

L. cylindrica ♀   x   L. acutangula ♂

↓

Hibrid F1

Figure 9. The hybridization plan

The research continued with the segregation of the hybrid made from the crossing of the two different species of Luffa. In F2 resulted 6 new genotypes with distinct features and numerous intermediate forms that have been removed in the process of breeding.

After intensive breeding works, these new genotypes were carefully selected and gene-tically stabilized in order to promote them in subsequent crops.

In what concerns the characteristics of the fruit in the green stage and as a sponger, the measurements showed significant differences between the studied genotypes (table 2).

Analyzing the fruit surface, G4 emphasized with the appearance of the protruding ribs on a smooth surface, unlike most genotypes that have exhibited a slightly rough and ribbed surface (Fig. 10).

Regarding the receptacle, the biggest diameter but also the greatest length was recorded by *L. acutangula,* having respectively 3,2 and 4,8 cm. Also the sponge color differs, having shades ranging from white, greenish white, yellowish and slightly brown.

In terms of sponge density, the genotypes with a small and medium density and by default a rare network are for fresh consumption, while those with high density cellulose fiber, and are intended for use as a vegetable sponge.

A dominant character was the black colour of the seeds which was transmitted in lineage for most of the genotypes, an exception being made only by *L. cylindrica* that has white seeds (Fig. 11).

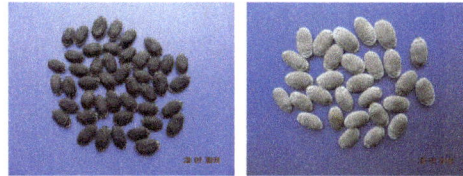

Figure 11. Black and white seeds

Observations were made regarding the fruit diameter in three stages: unripe, dried and as a sponge; the significant differences between the three stages were mainly due to dehydration of the fruit. (fig. 12).

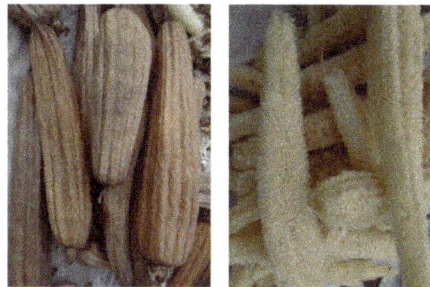

Figure 12. Dehydrated fruits and cellulose fiber sponge

Figure 10. Types of surfaces and fibers of the fruit

Table 2. Fruit characteristics

Variety name	Exterior color of the fruit	Fruit surface		Floral receptacle		Sponge colour	Sponge density	Seed colour
				Diameter (cm)	Length (cm)			
Luffa acutangula	Green	ribbed	Rough	3.2	4.8	Yellowish	Low	Black
Luffa cylindrica	Green	Slightly ribbed	Smooth	2.1	3.7	White	High	White
Luffa CxA	Green	Slightly ribbed	Slightly rough	2.7	4.1	Slightly brown	High	Black
G1	Green	Slightly ribbed	Rough	1.1	3.1	Slightly brown	Medium	Black
G 2	Green	Ribbed	Slightly rough	1.1	2.2	Greenish-white	High	Black
G 3	Light green	Ribbed	Smooth	1.4	4.1	Yellowish	High	Black
G 4	Light green	Heavily ribbed	Smooth	1.8	3.5	White	Low	Black
G 5	Light green	Ribbed	Smooth	0.8	2.8	White	Low	Black
G 6	Light green	Slightly ribbed	Slightly rough	2.5	4.2	Greenish-white	High	Black

Thereby, the upper part of the unripe fruit, the point of grip of the fruit from the stalk, has registered the largest diameter in the case of G2 with 6.6 cm reaching the stage of sponge to 5.1 cm. The greatest difference between the unripe fruit base diameter and the base diameter of sponge was recorded by G1 and G3, with a decreasing diameter of 1.7 cm from one stage to another.

In terms of the middle part of the fruit, the highest diameter value was registered at G2 with 9.9 cm for the unripe fruit and at the opposite boundary is G1 with 5.9 cm as a sponge. Regarding the apex diameter, it was found that G2 registered a maximal value of 11.5 cm as an unripe fruit and a minimal value of 6.7 cm as a dried pepo (Table no. 3).

Table 3. Unripe, dried and sponge fruit diameter

Variety name	Base (cm)			Middle (cm)			Apex (cm)		
	Unripe	Dried	Sponge	Unripe	Dried	Sponge	Unripe	Dried	Sponge
Luffa acutangula	6.1	5.5	4.9	8.9	8.4	7.8	10.8	10.2	8.7
Luffa cylindrica	5.5	4.6	4.0	8.3	7.8	7.1	8.9	8.3	7.6
Luffa CxA	6.3	5.6	4.7	9.5	8.9	8.4	10.9	10.1	9.3
G 1	5.6	4.7	3.9	7.1	6.6	5.9	9.4	8.6	7.9
G 2	6.6	5.8	5.1	9.9	9.6	8.8	11.5	10.8	10.1
G 3	5.6	4.5	3.9	8.9	8.2	7.7	9.7	8.9	8.3
G 4	5.3	4.5	4.0	9.6	9.0	8.4	10.4	9.5	8.9
G 5	3.7	2.8	2.2	7.8	7.0	6.5	7.5	6.7	6.8
G 6	6.3	5.4	4.8	9.0	8.5	7.8	11.1	10.4	9.8

CONCLUSIONS

At the present, researches finalised with the achievement of a solid germplasm collection both for L. cylindrica and L. acutangula.
For L. acutangula, we managed its acclimation, breeding and drafting of the crop technology.
Two new varieties were achieved with distinct destination use, L. cylindrica as a vegetable sponge and L. acutangula for fresh consumption.
A new hybrid was obtained from L. cylindrica X L. acutangula crossing, a hybrid that manifests the heterosis phenomenon with a high yield production, strength, genetic resistance at the main pathogens and a mixed use destination. The unripe fruits can be used in nutrition just like zucchini, and when they mature can be used as vegetable sponges because of the high density fibre, volume and their large water retention capacity.
After the achieved hybrid segregated, six new distinct genotypes were obtained, three of which can be used for producing vegetable sponges and three are for fresh consumption.

REFERENCES

Kousik, C. S., Levi, A., Wehner, T. C. and Maynard, D. N. 2015. Cucurbitaceae (Vine Crops). eLS. 1–8
Partap S., Kumar A., Sharma N.K., Jha K.K., 2012. Luffa Cylindrica: An important medicinal plant. J. Nat. Prod. Plant Resour, 2 (1): 127-134.
Prakash K., Pandey A., Radhamani J., Bisht I.S., 2013. "Morphological variability in cultivated and wild species of Luffa (Cucurbitaceae) from India." Genetic resources and crop evolution, vol 60, issue 8: 2319-2329.

PERMISSIONS

All chapters in this book were first published in SPSBH, by University of Agronomic Sciences and Veterinary Medicine of Bucharest, Romania; hereby published with permission under the Creative Commons Attribution License or equivalent. Every chapter published in this book has been scrutinized by our experts. Their significance has been extensively debated. The topics covered herein carry significant findings which will fuel the growth of the discipline. They may even be implemented as practical applications or may be referred to as a beginning point for another development.

The contributors of this book come from diverse backgrounds, making this book a truly international effort. This book will bring forth new frontiers with its revolutionizing research information and detailed analysis of the nascent developments around the world.

We would like to thank all the contributing authors for lending their expertise to make the book truly unique. They have played a crucial role in the development of this book. Without their invaluable contributions this book wouldn't have been possible. They have made vital efforts to compile up to date information on the varied aspects of this subject to make this book a valuable addition to the collection of many professionals and students.

This book was conceptualized with the vision of imparting up-to-date information and advanced data in this field. To ensure the same, a matchless editorial board was set up. Every individual on the board went through rigorous rounds of assessment to prove their worth. After which they invested a large part of their time researching and compiling the most relevant data for our readers.

The editorial board has been involved in producing this book since its inception. They have spent rigorous hours researching and exploring the diverse topics which have resulted in the successful publishing of this book. They have passed on their knowledge of decades through this book. To expedite this challenging task, the publisher supported the team at every step. A small team of assistant editors was also appointed to further simplify the editing procedure and attain best results for the readers.

Apart from the editorial board, the designing team has also invested a significant amount of their time in understanding the subject and creating the most relevant covers. They scrutinized every image to scout for the most suitable representation of the subject and create an appropriate cover for the book.

The publishing team has been an ardent support to the editorial, designing and production team. Their endless efforts to recruit the best for this project, has resulted in the accomplishment of this book. They are a veteran in the field of academics and their pool of knowledge is as vast as their experience in printing. Their expertise and guidance has proved useful at every step. Their uncompromising quality standards have made this book an exceptional effort. Their encouragement from time to time has been an inspiration for everyone.

The publisher and the editorial board hope that this book will prove to be a valuable piece of knowledge for researchers, students, practitioners and scholars across the globe.

LIST OF CONTRIBUTORS

Costel Vînătoru, Bianca Zamfir and Camelia Bratu
Vegetable Research and Development Station Buzău, No. 23, Mesteacănului Street, zip code 120024, Buzău, Romania

Victor Lăcătuş and Luminiţa Cârstea
Research and Development Institute for Vegetable and Flower Growing Vidra, 22 Calea Bucuresti Street, Bucharest, Romania

Constanţa Alexe, Marian Vintilă and Simona Popescu
Research and Development Institute for Processing and Marketing of the Horticultural Products – HORTING Bucharest, No. 1A, Intrarea Binelui Street, District 4, 042159, Bucharest, Romania

Gh. Lămureanu
Research Station for Fruit Growing (R.S.F.G) Constanta, No.1, Pepinierei Street, 907300, Commune Valu lui Traian, Romania

Lenuţa Chira
University of Agronomic Sciences and Veterinary Medicine of Bucharest, No. 59, Marasti Blvd, District 1, 011464, Bucharest, Romania

Aurelia Diaconu, Reta Drăgici, Mihaela Croitoru, Marieta Ploae, Iulian Drăghici and Milica Dima
Research - Development Center for Agricultural Plants on Sands, Dabuleni, Dolj County, Romania

Cho Eun-Gi
Kyungpook National University in South Korea

Leonard Boroş
Phytosanitary Unit, Regional Laboratory of Nematology, 47 Lânii Street, 500465, Braş ov, Romania

Tatiana Eugenia Şesan and Mariana Carmen Chifiriuc
University Bucharest, Faculty of Biology, 2aResearch Institute of the University of Bucharest – ICUB, Spl. Independenţei 91-95, Bucharest, Romania

Ionela Dobrin and Beatrice Iacomi
University of Agronomic Sciences and Veterinary Medicine of Bucharest, 59 Blvd. Mărăşti, Bucharest, 011464, Romania

Claudia Costache
Central Phytosanitary Laboratory, 11 Voluntari Blv.,077190 Voluntari, Ilfov, Romania

GüLtekin Özdemir
Dicle University, Faculty of Agriculture, Department of Horticulture, Diyarbakir, Turkey

Petre Marian Brezeanu, Silvica Ambarus, Tina Oana Cristea and Maria Calin
Vegetable Research and Development Station Bacău, Calea Bîrladului No 220, Bacău, Romania

Creola Brezeanu
Vegetable Research and Development Station Bacău, Calea Bîrladului No 220, Bacău, Romania
Ion Ionescu de la Brad" University of Agricultural Sciences and Veterinary Medicine Iasi, Faculty of Agriculture, 3, Mihail Sadoveanu Alley, Iasi, Romania

Teodor Robu
Ion Ionescu de la Brad" University of Agricultural Sciences and Veterinary Medicine Iasi, Faculty of Agriculture, 3, Mihail Sadoveanu Alley, Iasi, Romania

Dimka Haytova and Nikolina Schopova
Agricultural University, 12 Mendeleev str., 4000 Plovdiv, Bulgaria

Gheorghiţa Hoza and Bogdan Gheorghe Enescu
University of Agronomic Sciences and Veterinary Medicine of Bucharest, 59 Mărăşti Blvd, District 1, 011464, Bucharest, Romania

Alexandra Becherescu
Banat University of Agricultural Sciences and Veterinary Medicine Timisoara, Calea Aradului nr. 119, 300645 Timişoara, Jud. Timiş, România

Elena Panţer and Maria Pele
University of Agronomic Sciences and Veterinary Medicine of Bucharest, Faculty of Biotechnologies, 59 Mărăşti Blvd, District 1, 011464, Bucharest, Romania

Elena Maria Drăghici
University of Agronomic Sciences and Veterinary Medicine of Bucharest – Faculty of Horticulture, 59 Mărăşti Blvd, District 1, 011464, Bucharest, Romania

Nadejda Mihnea, Galina Lupaşcu and Irina Zamorzaeva
Academy of Sciences of Moldova, Institute of Genetics, Physiology and Plant Protection, 20 Pădurii Street, MD-2002, Chisinau, Republic of Moldova

Petre Sorin and Maria Pele
University of Agronomic Sciences and Veterinary Medicine of Bucharest – Faculty of Biotechnologies, 59 Mărăşti Blvd, District 1, 011464, Bucharest, Romania

Atilgan Atilgan and Hasan Oz
Suleyman Demirel University, Agriculture Faculty, Agricultural Structure and Irrigation Department, 32260 Isparta, Turkey

Ali Yucel
Osmaniye Korkut Ata University, Osmaniye Vocational School, 80000 Osmaniye, Turkey

Cagatay Tanriverdi
Kahramanmaras Sutcu Imam University, Agriculture Faculty, 46100 Kahramanmaraş, Turkey

Ahmet Tezcan
Akdeniz University, Agriculture Faculty, Agricultural Structure and Irrigation Department, 07100 Antalya, Turkey

Nikolina Shopova and Dimka Haytova
Agricultural University, 4000 Plovdiv, 12 "Mendeleev" Str. Bulgaria

Camelia Bratu, Costel Vînătoru, Bianca Zamfir and Elena Bărcanu
Vegetable Research and Development Station Buzău, No. 23, Mesteacănului Street, zip code 120024, Buzău, Romania

Florin Stănică and Viorica Lagunovschi
University of Agronomic Sciences and Veterinary Medicine of Bucharest, 59 Marasti Blvd., District 1, Bucharest, Romania

Senem Tülek
Ministry of Agriculture and Rural Affairs, Central Plant Protection Research Institute, 06172, Yenimahalle, Ankara, Turkey

Fatma Sara Dolar
Ankara University, Faculty of Agriculture, Department of Plant Protection, 06110, Ankara

Steliana Rodino, Marian Butu, Gina Fidler and Alina Butu
National Institute of Research and Development for Biological Sciences, Splaiul Independentei, no 296, 060031, Bucharest, Romania

Ancuţa Marin
Research Institute for Agricultural Economics and Rural Development, 61 Blvd. Marasti, 011464, Bucharest, Romania

Kenan Kaynaş
Çanakkale Onsekiz Mart University. Faculty of Agriculture. Terzioglu Campus. 17020. Çanakkale. Turkey

Marian Vintilă and Florin Adrian Niculescu
Research and Development Institute for Processing and Marketing of the Horticultural Products - Bucharest, No. 1A, Intrarea Binelui Street, District 4, 042159, Bucharest, Romania

Milena Yordanova
University of Forestry, Faculty of Agronomy, 10 Kliment Ohridski Blvd, 1756, Sofia, Bulgaria

Nina Gerasimova
Institute of Plant Physiology and Genetics, Bulgarian Academy of Sciences, Acad. G. Bonchev Street, Bldg. 21,1113, Sofia, Bulgaria

Mădălina Doltu, Marian Bogoescu and Dorin Sora
Research and Development Institute for Processing and Marketing of Horticultural Products – Horting, 1A Intrarea Binelui Street, District 4, 042159, Bucharest, Romania

Vlad Bunea
Central School, Bucharest, Romania, 3-5 Icoanei Street, District 2, 20451, Bucharest, Romania

Nilda Ersoy
Akdeniz University, Vocational School of Technical Sciences, Program of Organic Agriculture, 07058, Antalya, Turkey

Osman Tekinarslan and Elif Akçay Özgür
ECAS Inspection and Certification Organization, 07075, Antalya, Turkey

Ulaş Göktaş
PROANALİZ Laboratories Group, Antalya, Turkey

Ayşe Metin
ESAY Agriculture Consulting, 07310, Antalya, Turkey

Ramazan Göktürk
Akdeniz University, Faculty of Science, Department of Biology, 07058, Antalya, Turkey

GüRkan Kaya
Mehmet Akif Ersoy University, Faculty of Engineering and Architecture, 15030, Burdur, Turkey

Zuzana Poórová and Zuzana Vranayová
Technical University of Košice, Civil Engineering Faculty, Vysokoškolská 4 Košice 042 00, Slovakia

Elena Delian, Adrian Chira, Liliana Bădulescu and Lenuţa Chira
University of Agronomic Sciences and Veterinary Medicine of Bucharest, 59 Mărăşti Blvd, District 1, 011464, Bucharest, Romania

Aurelia Dobrescu, Gheorghiţa Hoza and Daniela Bălan
University of Agronomic Sciences and Veterinary Medicine Bucharest 59 Marasti Blvd., 11464 Bucharest, Romania

Elena Dobrin, Gabriela Luţă, Evelina Gherghina, Daniela Bălan and Elena Drăghici
University of Agronomic Sciences and Veterinary Medicine of Bucharest, 59 Marasti Blvd, District 1, Bucharest, Romania

Nadejda Mihnea
Institute of Genetics, Physiology and Plant Protection, Academy of Sciences of Moldova, 20 Pădurii street, MD-2002, Chişinău, Republic of Moldova

Vera Stefanova
Agricultural University of Plovdiv, 12 Mendeleev Blvd, Plovdiv, Bulgaria

Constanţa alexe and marian vintilă
Research and Development Institute for Processing and Marketing of the Horticultural Products – Bucharest- HORTING, No. 1A, Intrarea Binelui Street, District 4, 042159, Bucharest, Romania

Gheorghe lămureanu
Research Station for Fruit Growing (R.S.F.G) Constanta, No.1, Pepinierei Street, 907300, Commune Valu lui Traian, Romania

Adrian Chira and Lenuța Chira
University of Agronomic Sciences and Veterinary Medicine Bucharest, No. 59, Mărăști Blvd, District 1, 011464, Bucharest, Romania

Ana Emilia Cenușă, Gabriela Șovărel, Marcel Costache and Elena Bratu
Research and Development Institute for Vegetables and Flower Growing Calea București 22, Vidra, Ilfov, Romania

Marius Velea
Holland Farming Agro SRL, București, Drumul Osiei 74, sect. 6

Costel Vînătoru, Bianca Zamfir and Camelia Bratu
Vegetable Research and Development Station Buzău, No. 23, Mesteacănului Street, zip code 120024, Buzău, Romania

Liliana Bădulescu and Viorica Lagunovschi
University of Agronomic Sciences and Veterinary Medicine of Bucharest, 59 Marasti Blvd, District 1, Bucharest, Romania

Index